EXAM PREPARATION
Fire Officer I & II

EXAM PREPARATION
Fire Officer I & II

Douglas Cline

THOMSON
DELMAR LEARNING™

Australia Canada Mexico Singapore Spain United Kingdom United States

Exam Preparation for Fire Officer I & II
Douglas Cline

Vice President, Technology and Trades SBU:
Alar Elken

Editorial Director:
Sandy Clark

Acquisitions Editor:
Alison Weintraub

Development Editor:
Jennifer A. Thompson

Marketing Director:
Dave Garza

Channel Manager:
William Lawrensen

Marketing Coordinator:
Mark Pierro

Production Director:
Mary Ellen Black

Senior Production Manager:
Larry Main

Production Editor:
Thomas Stover

Editorial Assistant:
Maria Conto

Printed in the United States of America.
1 2 3 4 5 XXX 09 08 07 06 05

For more information contact:
Thomson Delmar Learning
Executive Woods
5 Maxwell Drive, PO Box 8007,
Clifton Park, NY 12065-8007
Or find us on the
World Wide Web
at www.delmarlearning.com

Library of Congress
Cataloging-in-Publication Data

1-4018-9922-6

Contents

Preface

Congratulations to you as you embark on the next step in your career in the fire service! Countless hours of work and training has led you to this point – studying for your certification exam. Whether you are a firefighter looking to step up to the role of officer, or whether you are an lieutenant aiming for a promotion to captain, the questions within these pages, as well as in the CD in the back of this book, will help prepare you for that final exam.

How To Use This Book

Based on the 2003 Edition of NFPA 1021 for Fire Officer training and Company Officer, 2E by Clinton Smoke, each exam was created to ensure that all officer competencies at levels I & II were covered. As such, each exam contains questions from the chapters in the book, as well as references to applicable NFPA Standards. In addition, because the questions within this book are also designed based on the NFPA Standard, the book will prepare you for any officer I and II exam.

The book is organized in a logical sequence based on Bloom's Taxonomy of learning, that progresses from test-taking strategies to questions about memorizing important terms and on to concepts and application of those concepts in a given scenario:

- The best place to start is at the introductory chapter, ***Acing Your Certification Exam.*** Here you will find valuable test-taking strategies, as well as advice on how to set up an effective study schedule prior to the certification exam date.
- ***Phase One: Comprehension*** contains three practice exams of 100 questions each to gauge your knowledge and ability to retain important facts and concepts.
- ***Phase Two: Application & Analysis*** contains three practice exams of 100 questions each that evaluate your ability to solve problems and recognize how important concepts fit together.
- ***Phase Three: Evaluation & Synthesis*** is the most complex stage, also containing three practice exams of 100 questions each, and judges how accurately you can make decisions based on given scenarios and your knowledge of the subject matter.
- ***Phase Four: Final Exam*** is a 200 questions test combining questions from all three phases, simulating the certification exam.
- ***Back of Book CD:*** contains *all* exams in a self-grading ExamView format, allowing you the option of self-study. Also included is a bonus final exam of 200 questions for additional practice in computer-based testing.
- ***Answers to Questions, NFPA and Textbook References and Rationale*** are provided for each question in all of the exams, allowing you to track your progress as well as the applicable sources for additional study as needed.

Successful candidates will take all of these exams in the order provided, using the answers, references, and rationale to grade their own work. The CD provides an excellent opportunity to retake tests as needed, as well as a bonus final exam. A copy of *Company Officer, 2E*, the reference material for the practice tests is also a useful tool for further study.

(Order #: 1-4018-2605-9)

Features of this Book

This book provides many features to enhance your learning experience and help lead you to closer to your goal of successful completion of the certification exam:

- ***Test Taking Strategies*** and ***Study Guidelines*** are outlined in the introductory chapter, allowing you to effectively practice, and fully prepare, for the exam.
- ***Questions of increasing complexity*** are organized in a logical sequence of the book to ensure accomplishment of important concepts related to the officer role.
- ***Answers to the Questions*** allow you to track your progress.
- ***NFPA References*** correlate the questions to the applicable NFPA 1021 Standard illustrating the competency which the question covers.
- ***Textbook References*** correlate the questions to the appropriate pages in the *Company Officer 2E* book, to allow for further study if needed.
- ***Rationale*** also accompanies each question, providing you with the necessary explanation to support the correct answer.
- ***Questions in ExamView format*** on the CD in the back of the book, allow you to use this book as a self-study tool and to practice computer-based testing. A bonus final exam is also provided.

Acknowledgements

The author and publisher gratefully acknowledge the content reviewers who participated in the project to ensure proper execution of this exam prep manual:

Mike Cox
Florida State Fire College
Ocala, FL

M.B. Oliver
Midland College
Midland, TX

William Shelton
Virginia Dept of Fire Programs
Richmond, VA

Gary Wilson
University of Missouri
Fire & Rescue Training Institute
Columbia, MO

About the Author

Douglas Cline is a 24-year veteran of the Fire Service serving as a Battalion Chief with the Chapel Hill Fire Department and an Adjunct Faculty for Great Oaks Institute of Technology Public Safety Services Division in Cincinnati, Ohio. He is a North Carolina Level II Fire Instructor, National Fire Academy Instructor for the Office of State Fire Marshal and an EMT-Paramedic instructor/coordinator for the North Carolina Office of Emergency Medical Services. Chief Cline was honored as the International Society of Fire Service Instructors, 1999 Instructor of the Year.

About the Series Advisor

Mike Finney is the head of the department of public safety services at Great Oaks Institute of Technology and Career Development. He has served on the NFPA 1041 committee in past years and is a current board member for the International Association of Fire Service Instructors. His area of expertise is teaching and curriculum development.

Acing the Certification Exam: An Introduction to Test-Taking Strategies

Introduction

Test time. Whether you are preparing for a certification test or a hiring test, the thought of an examination strikes fear in many people's hearts. The fear is so common that psychologists even have a diagnosis called test anxiety. However, testing does not have to be that way. Evaluations are simply an instrument to determine if you were effectively taught the information intended, or if you have the knowledge base necessary to do the job. That's all! If the purpose of testing is so simple, why do so many people become so anxious when test time comes. Several factors play into test anxiety and why so many people have such fears of testing. However, these can be overcome. With the assistance of this guide, you too can be better prepared and calmer on examination day.

Differences between hiring and certification tests

A hiring test is intended to determine if an applicant has the necessary knowledge base to perform the job. Different departments have different requirements for the job that is being filled, and those doing the hiring want to know that the applicant can meet the challenge. The outcome is very simple—get the most qualified applicant for the position. This can be the opportunity for you to "show your stuff." Hiring tests are designed to give each applicant an objective avenue to present their skill base. Careful preparation will give you the opportunity to excel. Every applicant has the same test, same questions, and same opportunity to prepare. As an applicant, take this opportunity to show the skills and knowledge base you have. Success on the hiring test is not necessarily measured on a set score or a "passing" score. Some departments set a cut-off score for applicants to proceed to the next phase, but many will take the top scores for the next phase. In this situation, it is imperative to get the highest score when compared to your competitors.

A certification test takes on a different dimension. Certification tests are designed to demonstrate mastery and are measured to national consensus standards. Since the other students do not measure the success, there is typically a set score that you must reach to be successful. The minimum score is set rather than being driven by all people taking the test.

How are test questions developed?

How questions are developed is critical to the understanding of how to take a test. Whether the examination is a hiring test, a promotional test, or a certification examination, they are developed around a clear set of objectives.

The learning objectives are a very important piece of the educational process. The objectives tell you several things:

- Under what conditions the student should be able to apply the knowledge. (condition)
- What type of knowledge the student should gain. (behavior)
- What depth or level of understanding the student should have. (standard)

While the order may vary based on the preference of the developer, this condition, behavior, and standard approach should be addressed in all learning objectives. When reviewing the learning objectives, take a moment to break it down into the individual pieces. This will provide insight into how the class should be taught. One will gain a wealth of information as to the intent of the designer by studying the learning objectives.

Domains of learning

Based on Bloom's Taxonomy, there are three primary areas or domains around which testing is designed. Knowing the domains will give an indication of the evaluation approach. The three domains of learning are:

- Cognitive domain- primarily deals with intellectual or knowledge skills, those that are intellectual processing or mental learning.
- Psychomotor domain- primarily deals with physical skills, those that require physically doing a task.
- Affective domain- primarily deals with a mindset, those that require a change in attitude or behavior.

Within the three domains were levels from the most basic understanding to advanced. Since we are primarily focusing on written tests, we will only deal with the cognitive domain. Table I-1 gives you the breakdown of the levels of understanding for the cognitive domain. The key to remember is that the higher the level you reach, the greater the understanding of the material must exist.

Table I-1

Knowledge: Recall of data	Examples: Recite a policy. Quote prices from memory to a customer. Knows the safety rules. Keywords: defines, describes, identifies, knows, labels, lists, matches, names, outlines, recalls, recognizes, reproduces, selects, states
Comprehension: Understand the meaning, translation, interpolation, and interpretation of instructions and problems. State a problem in one's own words.	Examples: Rewrites the principles of test writing. Explain in one's own words the steps for performing a complex task. Translates an equation into a computer spreadsheet. Keywords: comprehends, converts, defends, distinguishes, estimates, explains, extends, generalizes, gives examples, infers, interprets, paraphrases, predicts, rewrites, summarizes, translates
Application: Use a concept in a new situation or unprompted use of an abstraction. Applies what was learned in the classroom into novel situations in the workplace.	Examples: Use a manual to calculate an employee's vacation time. Apply laws of statistics to evaluate the reliability of a written test. Keywords: applies, changes, computes, constructs, demonstrates, discovers, manipulates, modifies, operates, predicts, prepares, produces, relates, shows, solves, uses
Analysis: Separates material or concepts into component parts so that its organizational structure may be understood. Distinguishes between facts and inferences.	Examples: Troubleshoot a piece of equipment by using logical deduction. Recognize logical fallacies in reasoning. Gathers information from a department and selects the required tasks for training. Keywords: analyzes, breaks down, compares, contrasts, diagrams, deconstructs, differentiates, discriminates, distinguishes, identifies, illustrates, infers, outlines, relates, selects, separates

Synthesis: Builds a structure or pattern from diverse elements. Put parts together to form a whole, with emphasis on creating a new meaning or structure.	Examples: Write a company operations or process manual. Design a machine to perform a specific task. Integrates training from several sources to solve a problem. Revises and processes to improve the outcome. Keywords: categorizes, combines, compiles, composes, creates, devises, designs, explains, generates, modifies, organizes, plans, rearranges, reconstructs, relates, reorganizes, revises, rewrites, summarizes, tells, writes
Evaluation: Make judgments about the value of ideas or materials.	Examples: Select the most effective solution. Hire the most qualified candidate. Explain and justify a new budget. Keywords: appraises, compares, concludes, contrasts, criticizes, critiques, defends, describes, discriminates, evaluates, explains, interprets, justifies, relates, summarizes, supports

Cognitive Domain (courtesy of http://www.nwlink.com/~donclark/hrd/bloom.html)

NFPA standards

The movement toward national standards for the fire service began in 1971 with the Joint Council of National Fire Service Organizations. The intent was to develop national performance standards. The end result, today, is 67 levels for 16 standards dealing with the professional qualifications for the fire service outlined by standards developed by the NFPA. Committees made up of training leaders, educators, private industry, and technical specialists guide standards development and revision. The intent is to give a better standard that meets a greater need. The standards are reviewed and revised as needed every five years. (This is why it is important when referencing an NFPA standard to also reference the year.) As a result, performance standards are kept up-to-date and accurate to what is needed in the field. While NFPA standards are not federal mandates, they do provide a widely accepted standard to follow. They provide the most current expectations for the position referenced.

The NFPA standards also provide an excellent reference for developing job descriptions. The 1000- series are referred to as the Professional Qualifications standards. Each provides a solid outline of performance requirements for almost any given position with the fire service. They are also easily adaptable to most any department. While there are other standards that can be followed, those set forth by the NFPA are the most widely recognized and are considered foundational for any department.

How does this apply to me?

NFPA standards are important to someone taking a fire-related examination, because often they are the basis for the learning objectives for test development. For certification examinations, NFPA standards are the foundation for many certifying bodies. For hiring and promotional tests, they are a solid basis for creating objective tests. A good resource to begin preparing for certification and hiring tests is the appropriate NFPA standard. Review the specific level and the requirements for such a level. This will give you some indication of what the test developer will be considering. As well, look at the requisite knowledge and skills. This will give you an indication of the extent of understanding of the information. An excellent way of reviewing the standards is to look at the verbs of the job performance requirement (JPR) in the context of Bloom's Taxonomy. Much like the requisite information, the JPR, as it relates to Bloom's Taxonomy, will tell you the level of understanding you must have with the information.

Test obstacles

Test obstacles are issues that complicate test taking. If we view test taking as simply an avenue to determine the individual's comprehension of the material, then test obstacles are barriers to the process. There are many issues that may create test obstacles. We will discuss a few.

Mental

Mental test obstacles can sometimes be the greatest hurdles to overcome. Mental preparation for a test can be as important as intellectual preparation. So often, many people have failed an exam before they even begin. Issues that arise out of mental obstacles are:

- feeling unprepared
- feeling incompetent
- fear of taking tests
- fear of failure

Overcoming these obstacles can be your greatest asset when testing. Not allowing yourself to be beaten before entering the testing area, can make the difference in success and failure on the exam.

Physical

Improper rest, poor eating habits, and lack of exercise can be some of the physical obstacles to overcome. When preparing for tests, always ensure that you get plenty of rest the night before, have a well-balanced meal before the test, and ensure you have a regiment of proper exercise. Physical obstacles are typically the easiest to overcome, however, the most overlooked.

Emotional

The emotional obstacles are often the most vague with which to deal. Much like the mental, emotional obstacles can cause a person to do poorly on an exam well before they enter the room. Stress-related issues that can interfere with test taking are:

- family concerns
- work-related concerns
- financial concerns

Emotional issues can cause a person to lose focus, cloud decision-making skills, and become a distracter. Overcoming these obstacles requires a conscious effort to ensure that the emotional does not interfere with the test.

Preparing to Take a Test

BEFORE the Test

1. Start preparing for the examination. For certification exams, start the first day of class. You can do this by reading your syllabus carefully to find out when your exams will be, how many there will be, and how much they are weighed into your grade. For hiring exams, it is recommended to begin studying at least eight weeks before the test.
2. For certification classes, plan reviews as part of your regular weekly study schedule; a significant amount of time should be used to review the entire material of the class.
3. Reviews are much more than reading and reviewing class assignments. You need to read over your class notes and ask yourself questions on the material you don't know well. (If your notes are relatively complete and well organized, you may find that very little rereading of the textbook for detail is needed.) You may want to create a study group for these reviews to reinforce your learning.

4. Review for several short periods rather than one long period. You will find that you are able retain information better and get less fatigued.
5. Turn the main points of each topic or heading into questions and check to see if the answers come to you quickly and correctly. Do not try to guess the types of questions; instead concentrate on understanding the material.

DURING the Test

1. Preview the test before you answer anything. This gets you thinking about the material. Make sure to note the point value of each question. This will give you some ideas on how best to allocate your time.
2. Quickly calculate how much time you should allow for each question. A general rule of thumb is that you should be able to answer 50 questions per hour. This averages out to one question every 1.2 seconds. However, make sure you clearly understand the amount of time you have to complete the test.
3. Read the directions CAREFULLY. (Can more than one answer be correct? Are you penalized for guessing? etc.) Never assume that you know what the directions say.
4. Answer the easy questions first. This will give you the confidence and give you a feel for the flow of the test. Only answer the ones for which you are sure of the correct answer.
5. Go back to the difficult questions. The questions you have answered so far may provide some indication of the answers.
6. Answer all questions (unless you are penalized for wrong answers).
7. Generally, once the test begins, the proctor can ONLY reread the question. He/she cannot provide any further information.
8. Circle key words in difficult questions. This will force you to focus on the central point.
9. Narrow your options on the question to two answers. Many times, a question will be worded with two answers that are obviously inaccurate, and two answers being close. (However, only one is correct.) If you can narrow your options to two, guessing may be easier. For example, if you have four options on a question, then you have a 25% chance of getting the question correct when guessing. If you can narrow the options to two answers, then you increase to a 50% chance of getting the correct choice.
10. Use all of the time allotted for the test. If you have extra time, review your answers for accuracy. However, be careful of making changes on questions of which you are not sure. Often times, people change the answer of questions of which they were not sure, when their first guess was correct.

AFTER the Test

Relax. The test has been turned in. You can spend hours second-guessing what you "could" have done, but the test is complete. For certification tests, follow up to see if you can find out what objectives you did well and what areas you could improve. Review your test if you can; otherwise, try to remap the areas of question and refocus your studying.

Preparation Plan

Once you have acquired the reference texts for the examination, begin by reviewing the introduction, the table of contents, and review how the book is organized. The introduction will tell you how the book has been set up and how it is intended to assist the individual with the learning process. The table of contents will provide a snapshot of how the book is organized. Scanning the text will give you an overview of the book's design. Once you have done this, break the chapters into four review sections. Using Table I-2 on the next page, fill in week 8 with the first section, week 7 with the second section, week 6 with the third section, and week 5 with the fourth section. Focus your energy into 50-minute increments with 10-minute breaks. Consider one hour for each chapter. However, on chapters in which you are competent, less time can be spent than with chapters that are more unfamiliar.

Week 4 will be spent taking section one of the Exam Prep guide. Base your time on 50 questions per hour. (100 questions should take 2 hours.) Do not check the answers until you have completed the entire test. Any questions missed should be reviewed and ensure you have an understanding of why the answer is correct.

Week 3 will be spent taking section two of the Exam Prep guide. Again, base your time on 50 questions per hour. At the end of the test, check answers and correct wrong answers.

Week 2 should be spent on section three of the Exam Prep guide. Maintain timeframes and check answers at the end of the test.

Week 1 should be spent with section four of the Exam Prep guide.

Five days before, go through the section one tests again.

Four days before, go through the section two tests again.

Three days before, go through the section three tests again.

Two days before, go through the section four tests again.

One day before do a light review of the text, focusing on areas you missed on the practice tests. However, take the evening and relax. Do something you enjoy, but make sure it is not a late night. Go to bed early and make sure you get a good night's sleep.

The day of the test make sure you have a well-balanced breakfast and arrive at the test site early.

Table I-2

PREPARATION GUIDE ***Plan based on a two-month schedule****	**STUDY NOTES**
Week 8 – Reference Text, Section I	
Week 7 –Reference Text, Section II	
Week 6 – Reference Text, Section III	
Week 5 – Reference Text, Section IV	
Week 4 – Exam Prep, Section I • Exam One • Exam Two • Exam Three	
Week 3 – Exam Prep, Section II • Exam One • Exam Two • Exam Three	
Week 2 – Exam Prep, Section III • Exam One • Exam Two • Exam Three	
Week 1 – Exam Prep, Section IV • Final Exam • Bonus Final Exam (on CD)	
5 days before – Review Section I Tests	

4 days before – Review Section II Tests	
3 days before - Review Section III Tests	
2 days before – Review Section IV Tests	
1 day before – Light review • Relax • Go to bed early	
Day of Test • Good breakfast • Arrive early	

Summary

Test taking does not have to be overwhelming. The obstacles to testing can be overcome and conquered through solid strategies and preparation. Initiating an effective plan, following it, and mentally preparing for a test can be your greatest tools to test success. As you work through the sections of this book, use the time as well, to work through some of the obstacles you face. When taking the tests in each section, try to simulate the environment of the actual test as much as possible. Successful testing is not an art. It is a learned skill. Through planning and practice anyone can acquire these skills.

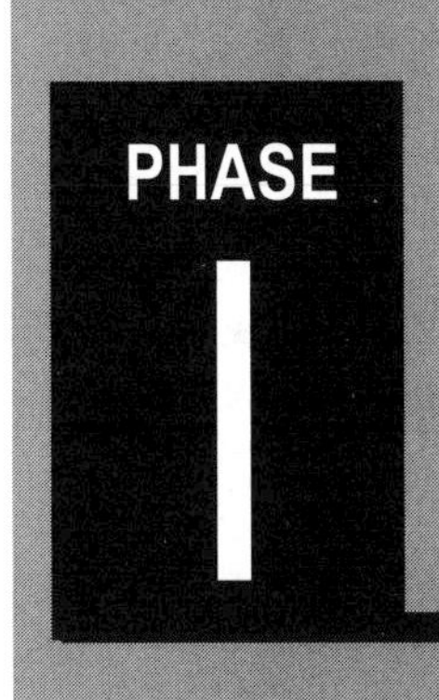

KNOWLEDGE & COMPREHENSION

Section one is designed to evaluate your basic understanding of the material. In this section, we are testing you understanding of definitions, recalling information, and identifying terms. Referring to Table I-1 (Bloom's Taxonomy, Cognitive Domain), we are covering the following levels:

- knowledge
- comprehension

Having mastered section one, you should be able to have a basic comprehension of the material.

PHASE I

Exam I

1. Actions taken during the first 50 minutes of an incident have a significant impact on the overall outcome of the incident.
   a. True
   b. False
2. Strong leadership and closed communications are vital to the success of any organization.
   a. True
   b. False
3. Lightweight truss construction is primarily used in Type II construction.
   a. True
   b. False
4. All of firefighters killed during structural firefighting operations were advancing hose lines at the time of their injury.
   a. True
   b. False
5. Persons with low affiliation needs don't like to work in groups.
   a. True
   b. False
6. Multi-company training evolutions provide an opportunity to exercise the Incident Management System.
   a. True
   b. False
7. Career development is a shared responsibility. Both the firefighter and the department have an obligation to this growth.
   a. True
   b. False
8. Persons with medium power needs like to be in charge.
   a. True
   b. False
9. Effective fire officers do not have the need to communicate effectively.
   a. True
   b. False

10. Regardless of whether the firefighter is a career employee or a volunteer member, injuries cost money.
    a. True
    b. False
11. In most departments, companies spend less than 90 percent of their time dealing with non-emergencies.
    a. True
    b. False
12. Ethics are an important management issue.
    a. True
    b. False
13. Planning does not have to start with a clear understanding of the goals and objectives.
    a. True
    b. False
14. Good time management correlates the management of our time with our previously set goals and objectives.
    a. True
    b. False
15. The crime of arson requires criminal intent as well as the act itself.
    a. True
    b. False
16. Leading others is the company officer's principle job.
    a. True
    b. False
17. Pre-incident planning is the fact-finding part of the pre-incident planning process.
    a. True
    b. False
18. The cycle of performance management represents the continuous process of goal setting, observation, and performance evaluation.
    a. True
    b. False
19. A fire department's Fire Suppression Rating Schedule is directly related to the rating classification given by the Insurance Service Office.
    a. True
    b. False
20. Your pre-incident planning should address the community's overall risk.
    a. True
    b. False

21. Good communication skills are not essential in both your work and personal life.
    a. True
    b. False

22. Experienced fire officers should be able to determine the origin and cause of nearly all the fires they attend.
    a. True
    b. False

23. Persons with high achievement needs like challenges.
    a. True
    b. False

24. One component in successfully introducing and managing change is to keep communications open.
    a. True
    b. False

25. An administrative process whereby an employee is punished for not conforming to the organizational rules or regulations is known as _______________.
    a. disciplinary action
    b. commendation
    c. recommendation
    d. citation

26. A formal dispute between employee and employer over some condition of employment is known as a ____________.
    a. grievance
    b. complaint
    c. conflict
    d. gripe

27. This duty involves activities that will identify hazards in occupancies.
    a. pre-plans
    b. pre-fire surveys
    c. inspections
    d. pre-incident analysis

28. A defined course of action is known as a(n) ______________.
    a. procedure
    b. policy
    c. action
    d. protocol

29. An effective tool in employee development that is described as helping a team member improve his knowledge, skills, and abilities is known as ___________________.
    a. counseling
    b. evaluation
    c. coaching
    d. all of the above

30. The purpose of the disciplinary process is to improve the subordinate's ___________.
    a. performance
    b. conduct
    c. both a & b
    d. none of the above

31. The hierarchy of needs was developed by ______________.
    a. Taylor
    b. Maslow
    c. Brunici
    d. Drucker

32. The fifth step in the management process involves the monitoring process to ensure that the work is accomplishing the intended objectives. It is known as _______________.
    a. controlling
    b. organizing
    c. coordinating
    d. commanding

33. What is defined as the personal actions of managers and supervisors to get subordinates to carry out certain actions?
    a. leadership
    b. direction
    c. discipline
    d. none of the above

34. The final step in the communications process is known as __________.
    a. feedback
    b. medium
    c. message
    d. receiver

35. Who was the father of professional management?
    a. Fayol
    b. Drucker
    c. Davis
    d. Reed

36. The five tiered representation of human needs developed by Abraham Maslow is known as __________.
    a. Maslow's Sequence of Needs
    b. Maslow's Directional Needs
    c. Maslow's Hierarchy of Needs
    d. Maslow's Hierarchy of Desires

37. A document that attests that one has demonstrated the knowledge and skills necessary to function in a particular craft and trade is a(n) __________.
    a. degree
    b. certification
    c. examination
    d. guideline

38. A company officer is what level of supervisor?
    a. up line
    b. down line
    c. first
    d. second

39. The activity of providing emergency services is known as a ____________.
    a. staff function
    b. non-essential function
    c. support function
    d. line function

40. The right and power to command is known as ______________.
    a. organizing
    b. planning
    c. management
    d. authority

41. The activity of getting things done through people is known as ____________.
    a. controlling
    b. planning
    c. management
    d. organizing

42. The NFPA Standard for Fire Officer is __________.
    a. 1001
    b. 1002
    c. 1021
    d. 1020

43. CIS stands for ____________.
    a. crew in service
    b. cat in sewer
    c. crazy inmate situation
    d. critical incident stress

44. The organizational principle that addresses the number of personnel a supervisor can effectively manage is referred to as ____________.
    a. command
    b. operations
    c. elements of behaviors
    d. span of control

45. Company officers are referred to as ____________.
    a. first-line supervisors
    b. up-line supervisors
    c. equals
    d. none of the above

46. The act of assigning duties to subordinates is known as __________.
    a. advocation
    b. delegation
    c. administration
    d. relocation

47. Arbitration is ______________.
    a. the process by which contract disputes are resolved with a facilitator
    b. a collective bargaining process
    c. an arrangement where employees are not required to join the union
    d. the process by which contract disputes are resolved with a decision by a third party

48. A system of values or standard of conduct is known as __________.
    a. defonics
    b. ethics
    c. procedures
    d. traditions

49. The act to disturb, torment, or pester means _____________.
    a. bug
    b. aggravate
    c. harassment
    d. none of the above

50. A financial plan to purchase high dollar items that have a life expectancy of more than one year is a _______________.
    a. line-item budget
    b. program budget
    c. capital budget
    d. operating budget

51. The easiest and most common form of communication is _____________.
    a. written
    b. oral
    c. radio
    d. email

52. A recognition of authority by virtue of the other individual's character or trust is __________ power.
    a. expert
    b. direct
    c. punishment
    d. identification

53. A target or other object by which achievement can be measured is known as a(n) ____________.
    a. objective
    b. goal
    c. note
    d. measure

54. The third step in the management process involves utilizing the talents of others while giving them direction to putting them into action, and is known as _______________.
    a. organizing
    b. planning
    c. commanding
    d. coordinating

55. A type of building construction in which the exterior walls are usually made of masonry and the interior of non-combustibles is known as Type ________________.
    a. I
    b. II
    c. III
    d. IV

56. The organizational concept that refers to the uninterrupted series of steps or layers within an organization is known as ___________.
    a. scaler organization
    b. unity of command
    c. linear organization
    d. flat organization

57. The expenses and possible income related to the delivery of a specific program within an organization is a __________ budget.
    a. line-item
    b. capital
    c. operating
    d. program

58. This object helps explain the ignition and extinguishment of fires.
    a. fire square
    b. fire triangle
    c. garden ohms
    d. fire tetrahedron

59. The concept of an organizational structure utilized on large emergency scenes whereby there is only one boss is known as __________.
    a. command
    b. operations
    c. unity of command
    d. division of labor

60. Which of the following is the deliberate and apparent process by which one focuses attention on the communications of another?
    a. oral communications
    b. written communications
    c. active listening
    d. passive listening

61. The providing of human resources in paid or volunteer departments is known as ____________.
    a. fire department
    b. public safety
    c. staffing
    d. jurisdiction

62. A formal statement that defines a course or method of action is known as __________.
    a. procedure
    b. protocol
    c. policy
    d. none of the above

63. Diversity is a quality of being ____________.
    a. diverse
    b. different
    c. all alike
    d. a & b
    e. b & c

64. Law or regulation that establishes minimum requirements for the design and construction of a building is __________.
    a. fire code
    b. inspections code
    c. building code
    d. permits

65. A financial plan to acquire the goods and services to run an organization for a specific period of time is a ____________ budget.
    a. capital
    b. operating
    c. line-item
    d. program

66. EEOC is an abbreviation for _______________.
    a. Equal Employment Opportunity Commission
    b. Equal Equine Opportunity Commission
    c. Equal Employment Opportunity Council
    d. Equal Employment Opportunity Committee

67. Company officers are _______________.
    a. up-line supervisors
    b. first-level supervisors
    c. third-level supervisors
    d. not supervisors at all

68. The father of Theory X and Theory Y management styles is ____________.
    a. Taylor
    b. McGregor
    c. Maslow
    d. Drucker

69. The act of being accountable for actions and activities along with having a moral and perhaps legal obligation to carry out certain activities is known as __________.
    a. accountability
    b. responsibility
    c. line of authority
    d. all of the above

70. The leadership style that is characterized by lots of direction and mostly one-way communications is known as __________.
    a. directing
    b. consulting
    c. supporting
    d. delegating

71. Officers being responsible for one's personal activities in the organizational context is known as __________.
    a. subordinate
    b. responsibility
    c. accountability
    d. line of authority

72. These factors keep people satisfied with their work environment.
    a. motivating factors
    b. hygiene factors
    c. responsibility factors
    d. none of the above

73. The leadership style where the supervisor essentially turns the management of the task over to subordinates is known as __________.
    a. directing
    b. consulting
    c. supporting
    d. delegating

74. A recognition of authority derived from the government or other appointing agency is __________ power.
    a. demanded
    b. granted
    c. legitimate
    d. reward

75. Total quality management or TQM was developed by __________.
    a. Maslow
    b. Ouchi
    c. Drucker
    d. Deming

76. Communications that tend to follow the customs, rules, and practices of the industry or work place are known as __________.
    a. formal communications
    b. informal communications
    c. oral
    d. documented

77. What is the leading cause of civilian fire deaths?
    a. cooking
    b. heating appliances
    c. children
    d. careless smoking
78. The type of building construction in which the structural components are noncombustible and protected from fire is known as __________.
    a. fire-resistive
    b. noncombustible
    c. heavy timber
    d. ordinary
79. The first step in the management process which involves looking into the future and determining objectives is known as __________.
    a. organizing
    b. planning
    c. strategies
    d. tactics
80. A rule for measuring or a model to be followed is __________.
    a. law
    b. policy
    c. document
    d. standard
81. The leading causes of fires include which of the following? 1. children playing with matches 2. cooking 3. electrical problems 4. heating 5. careless use of smoking materials
    a. 1, 3, 5 only
    b. 1, 2 ,4, 5 only
    c. all of the above
    d. none of the above
82. In the elements of communications process, the model starts with which of the following?
    a. feedback
    b. sender
    c. receiver
    d. medium
83. A systematic arrangement of a body of rules is __________.
    a. standard
    b. NFPA
    c. skills
    d. codes

84. A legal term denoting deliberate and unlawful burning of property is known as __________.

    a. fire
    b. arson
    c. incendiary devices
    d. abuse

85. A characteristic of organizational structures denoting the relationship between supervisors and subordinates is known as __________.

    a. line of authority
    b. accountability
    c. organization theory
    d. group dynamics

86. The arrangement of police, fire and other emergency response organizations placed into one organization is known as a __________ department.

    a. public safety
    b. combination
    c. professional
    d. public fire

87. The traditional organizational structure design where each individual member of the crew answers directly to the officer is known as __________ organizational structure.

    a. flat
    b. round
    c. scaler
    d. pyramid

88. The quality of combustible material in a given building is known as __________.

    a. fire load
    b. fuel load
    c. burn load
    d. fire power
    e. both a & b

89. The report of a study published in 1973 by the National Commission on Fire Prevention and control is known as __________.

    a. America Did It
    b. America Burning
    c. America Burning Revisited
    d. Councils Report on the Nations Fire Problem

90. The second step in the management process involves the bringing together and arranging the essential resources to get the job done. It is known as __________.
    a. strategies
    b. tactics
    c. planning
    d. organizing

91. Which of the following is an obstacle in the communications process?
    a. physical barrier
    b. personal barrier
    c. semantic barrier
    d. all of the above

92. The NFPA Health and Safety standard is numbered __________.
    a. 1400
    b. 1500
    c. 1403
    d. 1021

93. The method of transmission in the communications model is known as the_________.
    a. receiver
    b. message
    c. medium
    d. feedback

94. A person assigned as the manager of the department's health and safety program is the __________ officer.
    a. safety
    b. health
    c. health and safety
    d. division of life safety

95. The activities that support personnel providing emergency services is known as __________.
    a. line function
    b. staff function
    c. support role
    d. administrator's role

96. A group of people working together to accomplish a task is known as a __________.
    a. pool
    b. task force
    c. group
    d. organization

97. A labor relations term denoting that union membership is a condition of employment is an __________ shop.

    a. union
    b. closed
    c. open
    d. agency

98. A formal document indicating the focus and values for an organization is known as the __________.

    a. vision
    b. introduction
    c. mission statement
    d. rules and regulations

99. In the communications model, the information being sent to another is known as the __________.

    a. message
    b. feedback
    c. medium
    d. none of the above

100. The fourth step in the management process which involves the manager's directing and overseeing the efforts of others is __________.

    a. coordinating
    b. commanding
    c. organizing
    d. controlling

Phase I, Exam I: Answers to Questions

1. F
2. F
3. T
4. F
5. T
6. T
7. T
8. F
9. F
10. T
11. T
12. T
13. F
14. T
15. T
16. T
17. T
18. T
19. T
20. T
21. F
22. T
23. T
24. T
25. A
26. A
27. C
28. A
29. C
30. A
31. B
32. A
33. A
34. A
35. A
36. C
37. B
38. C
39. D
40. D
41. C
42. C
43. D
44. D
45. A
46. B
47. D
48. B
49. C
50. C
51. B
52. D
53. A
54. C
55. B
56. A
57. D
58. B
59. C
60. C
61. C
62. C
63. D
64. C
65. B
66. A
67. B
68. B
69. B
70. A
71. C
72. B
73. D
74. C
75. D
76. A
77. D
78. A
79. B
80. D
81. C
82. B
83. D
84. B
85. A
86. A
87. A
88. E
89. B
90. D
91. D
92. B
93. C
94. C
95. B
96. D
97. B
98. C
99. A
100. A

Phase I, Exam I: Rationale & References for Questions

Question #1. The initial action plan dictates the outcome based upon strategies set and actions or tactics deployed. NFPA 4.6, 5.6. *CO, 2E,* Page 320.

Question #2. Most structured departments have a pyramid style design that it is important to be able to communicate from top to bottom and bottom to top. NFPA 4.2. *CO, 2E,* Page 60.

Question #3. Lightweight construction is in Type II or Type V based on if it is wood or steel. NFPA 4.6.1. *CO, 2E,* Page 251.

Question #4. This is due to the fact most firefighters die early on in incidents and hose advancement is generally in early stages of incidents. This results in nearly half of the on-scene line of duty deaths. NFPA 4.7.2, 5.7.1. *CO, 2E,* Page 178.

Question #5. The fire service is typically a team organization in every aspect. NFPA 4.2.2, 5.2. *CO, 2E,* Page 125.

Question #6. ICS should be used during any and all incidents—training or real. NFPA 4.6, 4.7. *CO, 2E,* Page 308.

Question #7. The officer helps the firefighter develop based on the community needs. NFPA 2-2.1. *CO, 2E,* Page 13.

Question #8. Most officers have high power needs. NFPA 5.2. *CO, 2E,* Page 125.

Question #9. Being able to write is an important part of your professional career. NFPA 4.1.2. *CO, 2E,* Page 28.

Question #10. Each worker's compensation claim is money paid out. NFPA 5.4.2, 5.7.1. *CO, 2E,* Page 177.

Question #11. Most of the time, officers deal with other functions rather than emergency response. NFPA 1021, 4.2. *CO, 2E,* Page 5.

Question #12. Every person in the fire service should live by the fire service code of ethics. NFPA 4.3.2, 4.3.3. *CO, 2E,* Page 91.

Question #13. Planning is a systematic process and should be on-going, beginning with the end in mind. NFPA 5.2.1, 4.2.6, 4.2.1. *CO, 2E,* Page 99.

Question #14. It is important to be able to manage time to have good efficiency. NFPA 4.7, 5.7.1. *CO, 2E,* Page 105.

Question #15. Arson has to be targeted at someone. NFPA 4.5.1, 4.5.2, 5.5.2. *CO, 2E,* Page 271.

Question #16. Working with others to accomplish the mission of the department is the focus of the company officer. NFPA 5.2.1, 4.2.1. *CO, 2E,* Page 5.

Question #17. Taking a look at the potentials is the beginning stages to developing action plans. NFPA 4.6.1, 5.6.1. *CO, 2E,* Page 292.

Question #18. This is a cycle due to the achievements and progressional growth of personnel. NFPA 5.2.2. *CO, 2E,* Page 152.

Question #19. The fire officer helps meet all of the objectives and the grading of ISO. NFPA 5.1.1. *CO, 2E,* Page 86.

Question #20. Taking a look at the potentials is the beginning stages to developing pre-incident plans. NFPA 4.6.1, 5.6. *CO, 2E,* Page 299.

Question #21. Effective communications are a vital part of your professional life. NFPA 4.1.2. *CO, 2E,* Page 20.

Question #22. Most fires are not investigated by an investigator, but by the company officer. NFPA 4.5.1, 4.5.2, 5.5.2. *CO, 2E,* Page 276.

Question #23. Most officers enjoy the challenges with which they are faced because to get to the officer level, you usually are a high achiever. NFPA 5.2 , 4.2.2. *CO, 2E,* Page 125.

Question #24. With change comes confusion and uneasiness. If officers can keep communication open and flowing, there is an excellent chance that change will be easier. NFPA 4.2.2, 4.2.3, 5.4.1. *CO, 2E,* Page 166.

Question #25. You will be required to take disciplinary actions as an officer. It is important to use them as a tool for redirection of personnel onto the department's mission pathway. NFPA 5.2.2. *CO, 2E,* Page 160.

Question #26. This is an avenue for the employee to pursue when they feel it is warranted and that they are in the right. NFPA 5.2.2, 4.2. *CO, 2E,* Page 163.

Question #27. Inspections are a proactive approach to fire prevention and are a great customer service. NFPA 5.5.1. *CO, 2E,* Page 226.

Question #28. Procedures are set ways of doing tasks or performing duties. NFPA 4.2.3, 4.4.2, 4.2.6. *CO, 2E,* Page 84.

Question #29. We do a lot of this leadership style as employees begin to come into their own talents. NFPA 4.2.3, 5.2.2. *CO, 2E,* Page 147.

Question #30. You will be required to take disciplinary actions as an officer. It is important to use them as a tool for redirection of personnel onto the department's mission pathway. NFPA 5.2.1. *CO, 2E,* Page 160.

Question #31. It is important to understand managerial theories to understand why personnel act and do as they do. It helps in developing leadership styles. NFPA 4.2.6, 4.4.2. *CO, 2E,* Page 81.

Question #32. For a gear system to run, it must be synchronized. The controlling of personnel efforts is like a machine running with the employees being the gears. NFPA 4.4.2. *CO, 2E,* Page 79.

Question #33. You, as an officer, are the one that leads the work force. Leadership is essential. NFPA 5.6.1, 4.6.3. *CO, 2E,* Page 120.

Question #34. This ensures that the receiver understood what message was being sent. NFPA 4.1.2. *CO, 2E,* Page 21.

Question #35. Company officers swill use many management styles to accomplish objectives. NFPA 4.2.6. *CO, 2E,* Page 77.

Question #36. Human behavior is important in the leadership and management styles you choose as an officer. NFPA 4.2.4, 5.2.1. *CO, 2E,* Page 123.

Question #37. Certification is justification for pay and promotions along with the professionalism that goes with being an officer. NFPA 3.3.3. *CO, 2E,* Page 8.

Question #38. Company officers are the first supervisors in the pyramid structure organization. NFPA N/A. *CO, 2E,* Page 4.

Question #39. Operations personnel are called line personnel where as they refer to the analogy of military as we are engaged in battles with the responses we encounter. NFPA 4.2.6. *CO, 2E,* Page 63 & 64.

Question #40. We as officers have a lot of power of personnel and situations. It is important that we utilize this power and authority with ethics in mind. NFPA 4.2, 4.6, 5.6. *CO, 2E,* Page 98.

Question #41. Company officers are the first line of management utilizing resources to accomplish objectives. NFPA 4.2.6. *CO, 2E,* Page 75.

Question #42. This is the professional standard to which fire officers are trained and certified. NFPA 1021. *CO, 2E,* Page 9.

Question #43. The mental wellness of crews is important for efficient and productive work. NFPA 4.7. *CO, 2E,* Page 205 & 206.

Question #44. It is important to be able to manage effectively as a company officer. NFPA 4.2.6. *CO, 2E,* Page 67.

Question #45. Company officers are the first level of supervisor and this is where most employees look to for leadership. NFPA4.2.1. *CO, 2E,* Page 62.

Question #46. To accomplish the goals of the organization, it is important to utilize all of your resources. NFPA 4.6.3. *CO, 2E,* Page 67.

Question #47. Working in a union environment is a very different situation and will vary from union to union. It is important to understand how the union functions and its goals. NFPA 4.2, 5.2.1, 5.1.1. *CO, 2E,* Page 103.

Question #48. We usually have a feeling of guilt when we do something wrong. Ethics often pick up where laws leave off. NFPA 4.3.2. *CO, 2E,* Page 89.

Question #49. This is an EEOC directive. NFPA 5.2.1; 4.2.1. *CO, 2E,* Page 133.

Question #50. As company officers, we affect the budget in our every day activities. NFPA 4.4.3 5.4.2. *CO, 2E,* Page 103.

Question #51. Spoken word is the most common used method of communication. NFPA 4.1.2. *CO, 2E,* Page 22.

Question #52. We have this due to the uniform, badge, and title. NFPA 4.2 and 5.2. *CO, 2E,* Page 127.

Question #53. Goals and objectives help us identify where we are going. NFPA 4.2.1, 5.2.2. *CO, 2E,* Page 98.

Question #54. We command scenes that are emergency and non-emergency. Station duties are commanded by the officer. NFPA 4.6.3, 4.4.1. *CO, 2E,* Page 78.

Question #55. Building construction and fire behavior go hand in hand. It is important to understand how fire will behave in particular build construction styles. NFPA 4.6.1. *CO, 2E,* Page 248 & 249.

Question #56. We climb the ladder of rank in the fire service progressing upward. NFPA 4.4. *CO, 2E,* Page 66.

Question #57. For a program to exist, it must be funded unless it can be done without any expenditures. NFPA 4.4.3, 5.4.2. *CO, 2E,* Page 103.

Question #58. Fire behavior is essential in fire suppression and investigation. NFPA 4.5.1. *CO, 2E,* Page 224.

Question #59. Many organizations and groups will work together on scenes. It is important to have one leader who is supported by a great staff whereas the leader can make decisions and communicate them. NFPA 4.6.2 section B. *CO, 2E,* Page 66.

Question #60. Officers should exercise active listening to make sure they are getting the message. It builds respect with the employees. NFPA 4.2.2 section A. *CO, 2E,* Page 24.

Question #61. The most valuable resource in the fire service is personnel. NFPA 4.2.2. *CO, 2E,* Page 54.

Question #62. Policies give us directions to follow to remain consistent. NFPA 4.2.5. *CO, 2E,* Page 84.

Question #63. We work in diverse communities and it is essential to have a diverse work force. NFPA 4.2.1, 5.2.1. *CO, 2E,* Page 132.

Question #64. The building code and fire code go hand in hand. This is a way that a proactive approach can be taken in the developmental stages of a construction project to enhance fire safety. NFPA 5.5. *CO, 2E,* Page 228.

Question #65. Most officers are aware of the operating budget since this is the concept we use mostly in our personal lives. NFPA 4.4.3, 5.4.2. *CO, 2E,* Page 103.

Question #66. This organization sets a lot of the rules and standards we are required to follow. NFPA 4.2.1, 5.2.1. *CO, 2E,* Page 131.

Question #67. Company officers are the direct supervisors of the workers. NFPA 4.2. *CO, 2E,* Page 121.

Question #68. It is important to understand managerial theories to understand why personnel act and do as they do. It helps in developing leadership styles. NFPA 4.2.6, 4.4.2. *CO, 2E,* Page 81.

Question #69. We as officers must take responsibility for our actions to be responsible for the actions of others. NFPA 4.2.6 section A. *CO, 2E,* Page 62.

Question #70. Employees who are less experienced need this style of leadership. NFPA 4.2, 5.2. *CO, 2E,* Page 128.

Question #71. You now are held accountable for other's actions when you step to the officer level. This is the first level where you are responsible for others. NFPA 4.2.6. *CO, 2E,* Page 62.

Question #72. We all like a good working environment. As a company officer, you should help ensure that the hygiene factors are there. This will increase productivity and morale. NFPA 5.7.1 and 4.7.1. *CO, 2E,* Page 124 & 125.

Question #73. When you have the resources and talent to use, it is important to utilize them to accomplish set goals and the mission. NFPA 4.2, 5.2. *CO, 2E,* Page 128.

Question #74. We are given this legitimate power when we are promoted to the officer level by the governing body. NFPA 4.2, 5.2. *CO, 2E,* Page 127.

Question #75. It is important to understand managerial theories to understand why personnel act and do as they do. It helps in developing leadership styles. NFPA 4.2.6, 4.4.2. *CO, 2E,* Page 82.

Question #76. Good communications skills are essential in both work and personal life. NFPA 4.2. *CO, 2E,* Page 20.

Question #77. As officers, it is important to understand the causes of fires and fire fatalities so you can reinforce the materials delivered during fire safety programs and contacts. NFPA 4.5.1. *CO, 2E,* Page 222.

Question #78. Building construction and fire behavior go hand in hand. It is important to understand how fire will behave in particular building construction styles. NFPA 4.6.1. *CO, 2E,* Page 248 & 249.

Question #79. As a company officer, you have to envision what is in the future and plan for it. NFPA 4.4. *CO, 2E,* Page 77.

Question #80. Standards set the minimum at which organizations should function. NFPA 3.3.3. *CO, 2E,* Page 9.

Question #81. As company officers and firefighters, we work hard to prevent fires in a proactive approach to fire fighting. NFPA 4.5, 4.5.1, 4.5.2. *CO, 2E,* Page 221 & 222.

Question #82. The sender must formulate a thought and send it out to a receiver. NFPA 4.1.2. *CO, 2E,* Page 21.

Question #83. Standards set the lower limits for any business. NFPA 3.3.3. *CO, 2E,* Page 9.

Question #84. Arson is an area that must be proven in a court of law. Not every illegally set fire is arson. NFPA 4.5.1. *CO, 2E,* Page 222.

Question #85. As company officers, you have authority and you have the power to use it. NFPA 4.2. *CO, 2E,* Page 62.

Question #86. These organizations protect the citizens of the community in a variety of ways. NFPA 4.3.1 section A. *CO, 2E,* Page 54.

Question #87. This is where individual employees are equal in abilities and tasks. The span of control must be followed. NFPA 4.4.2. *CO, 2E,* Page 67 & 68.

Question #88. Building construction and fire behavior go hand in hand. It is important to understand how fire will behave in particular building construction styles. NFPA 5.6.1 4.6.2, 4.6.3. *CO, 2E,* Page 252.

Question #89. It is important to understand the fire problem in America. NFPA 5.5, 4.7. *CO, 2E,* Page 214.

Question #90. Organization skills are important and often one of the weakest links in the officer's toolbox. NFPA 4.4.1. *CO, 2E,* Page 78.

Question #91. Barriers can obscure messages. NFPA 4.2.1 section A. *CO, 2E,* Page 23 & 24.

Question #92. These are standards to which the fire service is held accountable. NFPA 1021 & 1500. *CO, 2E,* Page 183.

Question #93. The medium is important because this is where the message could be lost or not understood due to barriers. NFPA 4.1.2. *CO, 2E,* Page 21.

Question #94. This is a standard set out by NFPA 1500. NFPA 5.7 & 4.7. *CO, 2E,* Page 184.

Question #95. The staff are the ones who make the budgets and are resource personnel. NFPA 4.2.6. *CO, 2E,* Page 63 & 64.

Question #96. The fire department is a group of people working to accomplish a task. NFPA 4.4.2 and 5.2.1. *CO, 2E,* Page 119.

Question #97. Working in a union environment is a very different situation and will vary from union to union. It is important to understand how the union functions and its goals. NFPA 4.2, 5.2.1 5.1.1. *CO, 2E,* Page 102.

Question #98. The mission statement is the road map for the destination of the organization. NFPA 4.2. *CO, 2E,* Page 87.

Question #99. Good communications is essential in both work and personal life. NFPA 4.1.2. *CO, 2E,* Page 21.

Question #100. For a gear system to run, it must be synchronized. The controlling of personnel efforts is like a machine running with the employees being the gears. NFPA 4.4.2 4.6.2. *CO, 2E,* Page 78.

PHASE I

Exam II

1. The standards of ethics never change.
   a. True
   b. False
2. Planning should always start with a clear understanding of the policy and procedures.
   a. True
   b. False
3. Most of the department's expenses are in the capital budget.
   a. True
   b. False
4. The consulting style of leadership involves mostly one-way communications.
   a. True
   b. False
5. One of the HSO's duties is to investigate accidents.
   a. True
   b. False
6. Firefighters have a high rate of heart and respiratory disease and cancer.
   a. True
   b. False
7. Fire prevention is the action taken to control and extinguish a fire.
   a. True
   b. False
8. America Burning was published in 1983 by the National Commission on Fire Prevention and Control.
   a. True
   b. False
9. FEMA's mission is to reduce the loss of life and property and protect our nation's critical infrastructure from all types of hazards, through a comprehensive emergency management program of risk reduction, preparedness, response, and recovery.
   a. True
   b. False
10. Many fire departments have no overall strategy for dealing with fire prevention issues.
    a. True
    b. False

11. Ventilation can be described as the procedure necessary to effect the planned and systematic direction and removal of smoke, heat, and fire gases from within a structure.
    a. True
    b. False
12. Flashover is a dramatic event in a room fire that rapidly leads to full involvement of all combustible materials present.
    a. True
    b. False
13. A fire tetrahedron consists of a fourth side to the fire triangle. This is called a continuous "chain reaction."
    a. True
    b. False
14. Physical factors include geographical size, population, valuation, response time, and topography of the community.
    a. True
    b. False
15. The lighter weight structural component is far less vulnerable to failure when exposed to fire.
    a. True
    b. False
16. Size-up is defined as a mental assessment of a situation; gathering and analyzing information that is critical to the outcome of an event.
    a. True
    b. False
17. A company officer should call for a fire investigator after any event in which the officer in charge is unable to determine the cause.
    a. True
    b. False
18. Information needed in pre-incident planning should include building number, street address, occupancy load during day and night, water supply, and access to utility cutoffs.
    a. True
    b. False
19. The Emergency Management Institute is a division of FEMA.
    a. True
    b. False
20. Pre-incident plan information should include company assignments, predicted strategies, or exposures.
    a. True
    b. False

21. Fire extinguishment means to cut off the fire from further advance, and to keep it from spreading and doing any further damage.
    a. True
    b. False
22. Life safety is the first priority during all emergency operations that addresses the safety of occupants and emergency responders.
    a. True
    b. False
23. Life safety is the third priority of a fire officer on the scene of an emergency.
    a. True
    b. False
24. Life safety is the first priority of a fire officer on the scene of an emergency.
    a. True
    b. False
25. Incident stabilization is the third priority of a fire officer on the scene of an emergency.
    a. True
    b. False
26. We can prepare to make the transition from firefighter to officer with __________.
    a. study and effort
    b. practice and patience
    c. hard work
    d. physical abilities
27. To the fire chief and other senior officers, the company officer represents the ____________.
    a. firefighters
    b. union
    c. mission
    d. company
28. A shortcoming of too many officers is the lack of __________.
    a. experience
    b. education
    c. communication skills
    d. training
29. According to the text, the word ______ means that an individual has been tested by an accredited examination body on clearly identified material and found to meet a minimum standard.
    a. licensing
    b. certification
    c. qualified
    d. bonded

30. Lieutenants should be certified as a Fire Officer 1 in accordance with __________.
    a. NFPA 1002
    b. NFPA 1001
    c. NFPA 1021
    d. NFPA 1027

31. Although the company officer is often required to help with resolving the problem at hand, the officer's primary job is to ____________.
    a. lead
    b. council
    c. discipline
    d. coach

32. What is the principal job of the company officer?
    a. leading others
    b. extinguishing fire
    c. life safety
    d. incident command

33. Which of the following choices represents the order of the communication process?
    a. feedback, message, receiver, sender
    b. receiver, sender, feedback, message
    c. sender, message, receiver, feedback
    d. sender, receiver, message, feedback

34. Walls, distance, and background noise are examples of __________.
    a. physical barriers to effective communications
    b. things that provide for effective communications
    c. personal barriers to effective communications
    d. semantic barriers to effective communications

35. Which of the following is a barrier to effective communications?
    a. physical barrier
    b. personal barrier
    c. semantic barrier
    d. all of the above

36. An SOP is an example of what type of communication?
    a. formal
    b. informal
    c. closed
    d. open

37. What are the common communication barriers?
    a. emotional, judgmental, social
    b. selective, educational, social
    c. physical, personal, semantic
    d. emotional, physical, social

38. Most fire departments in America consist of ________ companies.
    a. paid
    b. volunteer
    c. social
    d. dedicated

39. The organizational principle whereby there is only one supervisor refers to ____________.
    a. unity of command
    b. chain of command
    c. scalar principle
    d. division of labor

40. According to the Company Officer textbook, the number of personnel that a supervisor can effectively manage is _____ to ______.
    a. 2, 4
    b. 4, 7
    c. 3, 5
    d. 5, 8

41. Firefighters informally evaluate their company officers on ___________.
    a. how the officers perform their duties
    b. how well the officers apply policy
    c. how fair the officers are among members
    d. all of the above

42. The organizational principle whereby there is a continuous chain of command refers to ____________.
    a. unity of command
    b. chain of command
    c. scalar principle
    d. division of labor

43. Span of control is ___________.
    a. delegation
    b. micro-management
    c. the organizational principle of one supervisor
    d. the number of people that can be effectively managed

44. The first step in the management process is _________.
    a. organizing
    b. coordinating
    c. planning
    d. commanding

45. ____________ is the breaking down of large tasks to smaller ones.
    a. Organizing
    b. Commanding
    c. Coordinating
    d. Controlling

46. Which describes the time focused on mid-range planning?
    a. one to five years
    b. within one year
    c. beyond five years
    d. none of the above

47. A system of conduct, principles of honor and morality, or guidelines for human actions describe ________.
    a. loyalty
    b. human rights
    c. ethics
    d. management

48. The accepted steps for solving problems are __________.
    a. gathering information
    b. accurately defining the problem
    c. looking for alternative solutions
    d. all of the above

49. Delegation of tasks is most difficult for ________________.
    a. chiefs
    b. first-time managers
    c. seasoned personnel
    d. probationary employees

50. The accepted steps for solving problems are __________.
    a. gathering information
    b. accurately defining the problem
    c. taking action
    d. monitoring results
    e. all of the above

51. A(n) __________ is a group of people working together to accomplish a task.
    a. union
    b. organization
    c. department
    d. club

52. The power bestowed upon an individual by the organization is referred to as ________ power.
    a. licensed
    b. legitimate
    c. legal
    d. legislative

53. The act of seeking advice or gathering information from another is known as _________.
    a. influencing
    b. consulting
    c. directing
    d. supporting

54. The first expression of dissatisfaction might be considered a _______________.
    a. conflict
    b. grievance
    c. complaint
    d. gripe

55. ___________ is an informational process that helps subordinates improve their skills and abilities.
    a. Counseling
    b. Coaching
    c. Evaluating
    d. Meeting

56. Three effective tools for employee development are ___________.
    a. education, opportunity, and advancement
    b. coaching, counseling, and evaluations
    c. education, counseling, and evaluations
    d. coaching, opportunities, and evaluations

57. The ____________ has a leading role in reinforcing safety.
    a. HSO
    b. fire chief
    c. company officer
    d. firefighter

58. According to NFPA 1500, every fire department should have an individual assigned to the duties of ______________ officer.
    a. supply clerk
    b. health and safety
    c. public education
    d. community relations

59. Each year, NFPA publishes ________ reports on firefighter safety.
    a. 2
    b. 4
    c. 6
    d. 8

60. Which one of the following is not part of the five generally accepted tools for dealing with critical incident stress (CIS)?
    a. training
    b. peer support
    c. counseling
    d. vacation

61. The ___________ has a leading role in reinforcing safety according to NFPA 1500.
    a. HSO
    b. fire chief
    c. company officer
    d. firefighter

62. Action taken to control and extinguish a fire is called __________.
    a. fire control
    b. fire prevention
    c. fire inspection
    d. fire suppression

63. __________ consists of two types of actions, namely fire suppression and fire prevention.
    a. Fire prevention
    b. Fire inspection
    c. Fire protection
    d. Fire service

64. Which of the following is not a cause of fire?
    a. electrical
    b. children
    c. elderly
    d. smoking

65. Control of fire growth is one of the elements of ___________.
    a. education
    b. engineering
    c. enforcement
    d. engagement

66. _______ is the leading cause of fires in residences and is by far the leading cause of fire-related injuries in residences.
    a. Arson
    b. Cooking
    c. Electrical
    d. A child

67. As defined, ____________ factors are an assessment of safety hazards for both civilians and firefighters in a particular occupancy which include stairwells and other penetrations to allow for rescue, fire spread, and potential falling hazards.
    a. resource
    b. access
    c. structural
    d. survival

68. Type ______ construction is called "fire resistive."
    a. I
    b. II
    c. III
    d. IV

69. The expected maximum amount of combustible material in a given fire area is the definition for __________.
    a. fire load
    b. combustible load
    c. fuel load
    d. smoke load

70. As defined, _____________ are an assessment of the consequences on the community, which includes the people, their property, and the environment.
    a. access factors
    b. physical factors
    c. occupancy factors
    d. community consequences

71. What are the three stages of fire growth within an enclosed structure?
    a. incipient phase, thermal phase, and smoldering phase
    b. free-burning phase, thermal phase, and smoldering phase
    c. incipient phase, free-burning phase, and smoldering phase
    d. smoldering phase, incipient phase, and thermal phase

72. The _____________ is the heat-absorbing capacity of a substance.
    a. specific heat
    b. British thermal unit
    c. latent heat of vaporization
    d. heat of combustion

73. The ________________ is the amount of heat required to convert a substance from a liquid to a vapor.
    a. specific heat
    b. British thermal unit
    c. latent heat of vaporization
    d. heat of combustion

74. First responder's observations _____________ can greatly help a fire investigator during an investigation and should be documented.
    a. during initial notification
    b. while en route to incident
    c. upon arrival
    d. all of the above

75. A(n) ______________________ is a written order issued by a court specifying the place to be searched and the reason for the search.
    a. administrative search warrant
    b. NFIRS report
    c. consent to search form
    d. criminal search warrant

76. There are three methods by which the fire department may reenter the property after the fire is extinguished. They include all of the following except __________.
    a. owner's consent to search form
    b. administrative search warrant
    c. probable cause
    d. search warrant

77. Common natural causes of fires are ______________ .
    a. lightning
    b. electricity
    c. smoking
    d. all of the above

78. Common natural causes of fires are ______________ .
    a. autoignition through chemical reaction
    b. electricity
    c. lightning
    d. answer (a) and (c)
    e. answer (b) and (c)

79. ___________ is defined as opened to the atmosphere by the fire burning through windows or walls through which head and fire by-products are released, and through which fresh air may enter.
    a. Oxidation
    b. Arson
    c. Vented
    d. Ventilation
80. A body of law that creates public regulatory agencies and defines their powers and duties is called __________.
    a. administrative search warrant
    b. administrative process
    c. search warrant
    d. probable cause
81. A fire investigator's written report should include __________.
    a. photos
    b. facts
    c. written notes
    d. all of the above
82. __________ is defined as a chemical reaction in which oxygen combines with other substances causing fire, explosions, and rust.
    a. Oxidation
    b. Energy
    c. Fuel
    d. Chemical reaction
83. Common causes of fires are ______________ .
    a. natural causes
    b. accidental
    c. arson
    d. all of the above
84. Common accidental causes of fires are ______________ .
    a. cooking
    b. electricity
    c. smoking
    d. heating equipment
    e. all of the above
85. Common causes of fire include all of the following except __________.
    a. cooking
    b. heating
    c. smoking
    d. arson

86. What is defined as activities associated with confining and extinguishing a fire?
    a. life safety
    b. fire control
    c. property conservation
    d. area of refuge

87. Pre-incident plans should address all of the following except __________.
    a. water supplies
    b. installed fire protection systems
    c. training
    d. apparatus placement and strategies

88. A __________ is a bird's eye view of a property showing existing structures for the purpose of pre-emergency planning, such as primary access points, barriers to access, utilities, and water supply.
    a. plot plan
    b. floor plan
    c. blueprint
    d. format layout

89. The first priority during all emergency operations that addresses the safety of occupants and emergency responders is __________ .
    a. fire control
    b. life safety
    c. property conservation
    d. access

90. ___________ is an activity in comprehensive emergency management.
    a. Mitigation
    b. Recovery
    c. Response
    d. Preparedness
    e. All of the above

91. What is the definition of transitional mode?
    a. firefighting operations that make a direct attack on a fire for purposes of control and extinguishment
    b. actions intended to control a fire by limiting its spread to a defined area
    c. the critical process of shifting from the offensive to the defensive mode or from the defensive to offensive
    d. none of the above

92. What is the definition of offensive mode?
    a. firefighting operations that make a direct attack on a fire for purposes of control and extinguishment
    b. actions intended to control a fire by limiting its spread to a defined area
    c. the critical process of shifting from the offensive to the defensive mode or from the defensive to offensive
    d. none of the above

93. As defined, tactics are ______________________________ .
    a. the firefighting operations that make a direct attack on a fire for purposes of control and extinguishment
    b. the duties and activities performed by individuals, companies, or teams
    c. a broad set of goals and outlines for the overall plan to control the fire
    d. the various maneuvers that can be used to achieve a strategy while fighting a fire or dealing with a similar emergency
    e. none of the above

94. What are the three operational modes?
    a. offensive, defensive, and command
    b. attack, defensive, and command
    c. offensive, defensive, and transitional
    d. none of the above

95. When transmitting "benchmarks" to the command for incident stabilization, the terminology used should be ____________ .
    a. "all clear"
    b. "under control"
    c. "loss stopped"
    d. "all OK"

96. For what does the "T" in "COAL WAS WEALTH" stand?
    a. time
    b. team
    c. temperature
    d. total

97. What is a defensive operational mode?
    a. establish command mode
    b. direct attack to confine and extinguish fire
    c. limiting the fire spread to a defined area
    d. protect exposures
    e. answer (c) and (d)

98. NFPA _______ is the "Standard on Emergency Services Incident Management System."
    a. 1651
    b. 1403
    c. 1561
    d. 1500

99. The ___________________ section is responsible for obtaining resources for an event.
    a. Operations
    b. Planning
    c. Logistics
    d. Finance/Administration
    e. None of the above

100. The Incident Action Plan is _______________ .
    a. a vivid but brief description of the on-scene conditions relevant to the emergency
    b. a list of assistance to be provided by another fire department or agency
    c. an organized course of action that addresses all phases of incident command within a specific period of time
    d. none of the above

Phase I, Exam II: Answers to Questions

1.	F	26.	A	51.	B	76.	C
2.	F	27.	D	52.	B	77.	A
3.	F	28.	C	53.	B	78.	D
4.	F	29.	B	54.	D	79.	C
5.	T	30.	C	55.	B	80.	B
6.	T	31.	A	56.	B	81.	D
7.	F	32.	A	57.	C	82.	A
8.	F	33.	C	58.	B	83.	D
9.	T	34.	A	59.	A	84.	E
10.	T	35.	D	60.	D	85.	D
11.	T	36.	A	61.	A	86.	B
12.	T	37.	C	62.	D	87.	C
13.	T	38.	B	63.	C	88.	A
14.	T	39.	A	64.	C	89.	B
15.	F	40.	B	65.	B	90.	E
16.	T	41.	D	66.	B	91.	C
17.	T	42.	C	67.	D	92.	A
18.	T	43.	D	68.	A	93.	D
19.	T	44.	C	69.	C	94.	C
20.	F	45.	A	70.	D	95.	B
21.	F	46.	A	71.	C	96.	A
22.	T	47.	C	72.	A	97.	E
23.	F	48.	D	73.	C	98.	C
24.	T	49.	B	74.	D	99.	C
25.	F	50.	E	75.	A	100.	C

Phase I, Exam II: Rationale & References for Questions

Question #1. The standards of ethics change. As officers, you have an obligation to act in an ethical manner at all times. NFPA 1021 4.1.1. *CO, 2E,* Page 90.

Question #2. Planning is a systematic process and should be ongoing. Planning should always start with a clear understanding of the goals and objectives. NFPA 1021 4.2.6; 5.4.2. *CO, 2E,* Page 99.

Question #3. Capital budgets are your big ticket items. However, the majority of your expenses come under your operating budget, which covers most everything else not covered in the capital budget. NFPA 1021 4.4.3; 5.4.2. *CO, 2E,* Page 103.

Question #4. In this style, there is some discussion with the supervisor by which they seek ideas, explain the needs and decisions, and sell the idea. NFPA 1021 4.2.2; 5.2. *CO, 2E,* Page 128.

Question #5. These duties are set forth in the NFPA 1500 and 1521 standards. NFPA 1021 4.7; 5.7. *CO, 2E,* Page 185.

Question #6. Approximately 40-50 percent of on-duty firefighter fatalities is related to heart attacks. Exposure to products and smoking is a key contributor to cancer. Many states and organizations are going to be signing a tobacco-free affidavit for all new employees. NFPA 1021 4.7; 5.7. *CO, 2E,* Page 203.

Question #7. Fire prevention is a proactive approach that involves the activities that help keep fires from occurring. NFPA 1021 4.6.1; 5.5. *CO, 2E,* Page 214.

Question #8. America Burning was published in 1973 and it was revisited in 1987. NFPA 1021 4.1.1; 5.5. *CO, 2E,* Page 214.

Question #9. This is accomplished through many disaster relief programs and agencies that support the fire service to include the National Fire Academy and the Fire Act Grants. NFPA 1021 5.1.1. *CO, 2E,* Page 232.

Question #10. This is true due to the departments not understanding the fire problems in their communities. NFPA 1021 4.3. *CO, 2E,* Page 222.

Question #11. Ventilation alters most of the fire-related phenomena and reduces the probability of backdraft and flashover. NFPA 1021 - 4.6 AND 5.6. *CO, 2E,* Page 261.

Question #12. Someone in a room at the flashover moment has little chance of survival. NFPA 1021 - 4.6 AND 5.6. *CO, 2E,* Page 258.

Question #13. This process describes the flaming or burning process where as the triangle describes the smoldering process. NFPA 1021 - 4.6 AND 5.6. *CO, 2E,* Page 257.

Question #14. These are risk factors and community consequences. These are important to understanding the communities fire problem as it relates to the construction features of the buildings within the district. NFPA 1021 - 5.5. *CO, 2E,* Page 248.

Question #15. Most lightweight construction has a failure point at about the 20-minute mark, which relates to the time line for fire operations to being in the early suppression activities. NFPA 1021 - 4.6. *CO, 2E,* Page 251.

Question #16. Size-up is critical in the set up of an operation. The tactics that are used to mitigate an incident are determined based upon size-up. This process of gathering and analyzing is an on-going process throughout the incident. NFPA 1021 - 4.6.2. *CO, 2E,* Page 265.

Question #17. If in doubt, ask for the fire investigator and ask as soon as possible. NFPA 1021 - 4.5 AND 5.5.2. *CO, 2E,* Page 277.

Question #18. Pre-planning is the process of preparing a plan for emergency operations at a given building or hazard. This effort will enhance the operations of incident mitigation. Not every incident can be planned for, but you can collect critical data that will help in the decision making and size-up if an incident occurs. NFPA 1021 4.6 AND 5.6. *CO, 2E,* Page 300.

Question #19. This division is located at the National Fire Academy in Emmittsburg, Md. NPFA 1021 - 4.6 and 5.6. *CO, 2E,* Page 290.

Question #20. In addition to events that occur at structures, departments should be prepared to deal with other situations that are likely to occur within their jurisdiction. NFPA 1021 4.2.3, 4.6 AND 5.6. *CO, 2E,* Page 303.

Question #21. Fire extinguishment means to extinguish all visible fire. This may mean that all of the fire is not extinguished and will require overhaul. NFPA 1021 4.2.3, 4.6, 4.7.1 AND 5.6. *CO, 2E,* Page 335.

Question #22. Life safety is the first tactical priority on any response. It starts with the life safety of the responders and is followed by the life safety of civilians. NFPA 1021 - 4.6 AND 5.6. *CO, 2E,* Page 314.

Question #23. Life safety is the first tactical priority on any response. It starts with the life safety of the responders and is followed by the life safety of civilians. NFPA 1021 - 4.2.1, 4.6 AND 5.6.1. *CO, 2E,* Page 320.

Question #24. Life safety is the first tactical priority on any response. It starts with the life safety of the responders and is followed by the life safety of civilians. NFPA 1021 - 4.2.1, 4.6 AND 5.6.1. *CO, 2E,* Page 320.

Question #25. You cannot be held accountable for what has happened prior to your arrival, but you can be held accountable for what happens after you arrive. You must be sure that your information, your planning, and your actions all support an effective, coordinated, and safe operation. NFPA 1021 - 4.2.1, 4.6 AND 5.6.1. *CO, 2E,* Page 320.

Question #26. It is important to gain as much practical and field knowledge as possible prior to becoming an officer. The transition is one that has significant impact on others within the organization. NFPA 1021 4.1.1. *CO, 2E,* Page 12.

Question #27. You are the leader of a group of individuals which make up a group. In the fire service, these individuals are not looked at most often as that, individuals. The group, however, is looked at as a whole and you as the leader of the group. Your success is dictated by the success of your company. NFPA 1021 4.2.5; 5.2.1. *CO, 2E,* Page 7.

Question #28. Communication skills are important in every aspect of an individual's life. NFPA 4.1.1. *CO, 2E,* Page 14.

Question #29. Certification provides a yardstick by which to measure competency in every type of department from the largest to the smallest and from all paid to all volunteer. NFPA 1021 4.1.1. *CO, 2E,* Page 8.

Question #30. This NFPA standard sets out the job performance requirements that set a competency level. NFPA 1021 4.1. *CO, 2E,* Page 13.

Question #31. The capabilities, efficiency, and morale of the company are direct reflections of the company officer's leadership ability. NFPA 1021 4.2. *CO, 2E,* Page 5.

Question #32. The capabilities, efficiency, and morale of the company are direct reflections of the company officer's leadership ability. NFPA 1021 4.2; 5.6. *CO, 2E,* Page 5.

Question #33. Being able to communicate effectively is important for both you and your department. Understanding the communications model is essential. NFPA 1021 4.2.5; 5.4.3. *CO, 2E,* Page 21.

Question #34. Barriers can cause ineffective communications. This can be critical especially on emergency scenes. NFPA 1021 4.2.1-4.3; 5.2.1- 5.4.5. *CO, 2E,* Page 23.

Question #35. Understanding what causes barriers and being able to adapt or eliminate these barriers will help in increasing the efficiency in your communications. NFPA 1021 4.2.5; 5.4.3. *CO, 2E,* Pages 23-24.

Question #36. SOPs are used as a formal communication as they set plans that are consistent across the department. NFPA 1021 4.2; 5.4.1. *CO, 2E,* Page 20.

Question #37. Understanding what causes barriers and being able to adapt or eliminate these barriers will help in increasing the efficiency in your communications. NFPA 1021 4.2.5; 5.6.2. *CO, 2E,* Pages 23-24.

Question #38. Due to the rural nature of much of the United States, the population of firefighters is volunteer-based upon the economical designs of communities and the need for responses. The larger the population, the more the demographics of the jurisdiction demand more service. NFPA 1021 4.1.1; 5.1.1. *CO, 2E,* Page 54.

Question #39. Unity of command is an essential organizational concept where employees or subordinates report to only one boss. NNFPA 1021 4.1.1; 5.1.1. *CO, 2E,* Page 66.

Question #40. To ensure that business is carried forward, it is important to have control of a group. The span of control is directly related to many factors like task and the leader's ability to supervise. NFPA 1021 4.2; 5.2. *CO, 2E,* Page 67.

Question #41. They do this based upon the officer's ability to perform his/her job, apply policy, and how fair they are. NFPA 4.2; 5.2. *CO, 2E,* Page 65.

Question #42. Unity of command is an essential organizational concept where employees or subordinates report to only one boss. NNFPA 1021 4.1.1; 5.1.1. *CO, 2E,* Page 66.

Question #43. To ensure that business is carried forward, it is important to have control of a group. The span of control is directly related to many factors like task and the leader's ability to supervise. NFPA 1021 4.2; 5.6. *CO, 2E,* Page 67.

Question #44. Planning is largely a mental process and often requires the ability to envision things that have not happened yet. NFPA 1021 4.2; 5.4. *CO, 2E,* Page 77.

Question #45. Breaking down of tasks to manageable levels allows progress to be made and measured easier than as a whole. NFPA 1021 4.2.6. *CO, 2E,* Page 78.

Question #46. Looking into the near future is important especially as you plan for budgets and projects. NFPA 1021 4.6.2; 5.4.2. *CO, 2E,* Page 77.

Question #47. Even where a professional organization sets standards for its individual members, there is always some degree of variance in how these rules are to be interpreted, and there are always a few who try to beat the systems. Ethics are trying to do what is right. NFPA 1021 4.1.1. *CO, 2E,* Page 89.

Question #48. When you are faced with solving a problem, you should follow a proven, logical approach to find a solution. NFPA 1021 4.2.4; 5.2.1. *CO, 2E,* Page 108.

Question #49. This is because many of these individuals have not had the opportunity to delegate work previously. NFPA 1021 4.2; 5.6.1. *CO, 2E,* Page 100.

Question #50. When you are faced with solving a problem, you should follow a proven, logical approach to find a solution. NFPA 1021 4.2.4; 5.2.1. *CO, 2E,* Page 108.

Question #51. Fire departments, companies and unions are organizations. It is important to understand how organizations work and function to be able to work effectively within them. NFPA 1021 4.1.1; 5.1.1. *CO, 2E,* Page 119.

Question #52. This is the power given to you because of the position. This creates demanded respect from individuals because the organization requires it. Company officers must earn respect for their abilities. NFPA 1021 4.1.1; 5.2. *CO, 2E,* Page 127.

Question #53. In this style of leadership, the officer still remains in close contact with the individual and gives a lot of direction. NFPA 1021 4.2.2; 5.2. *CO, 2E,* Page 128.

Question #54. Gripes and complaints need to be channeled into a positive action. NFPA 1021 4.2.4; 5.2.1. *CO, 2E,* Page 163.

Question #55. A company officer is much like a coach working with the student until the desired level of competence is demonstrated. NFPA 1021 4.2.4; 5.7. *CO, 2E,* Page 147.

Question #56. Utilizing these tools will help you as a supervisor develop students to a new level. NFPA 1021 4.2; 5.2.2. *CO, 2E,* Page 142.

Question #57. The constant interaction between firefighters and company officers puts the officer in a position to be a strong advocate for safety. NFPA 1021 4.7; 5.7. *CO, 2E,* Page 208.

Question #58. This is information and objectives set forth by the NFPA standard. NFPA 1021 4.7; 5.7. *CO, 2E,* Page 184.

Question #59. One report is on firefighter injuries and the other is on firefighter deaths. NFPA 1021 4.7; 5.7. *CO, 2E,* Page 179.

Question #60. A company officer spends a lot of time with the firefighter. They should be able to recognize signs and symptoms of critical incident stress and be able to help with the defusing of a situation offering sound advice on how to handle CIS. NFPA 1021 4.2.4; 5.2.1. *CO, 2E,* Page 206.

Question #61. This is information set forth by the NFPA 1500. NFPA 1021 4.7; 5.7. *CO, 2E,* Page 208.

Question #62. Fire suppression is denoted as a reaction with resources mobilized after an event. NFPA 1021 4.2.1. *CO, 2E,* Page 214.

Question #63. Understanding both types of fire protection is important along with understanding the community's fire problem. NFPA 1021 4.1.1. *CO, 2E,* Page 214.

Question #64. These are directed to human behavior. Elderly have a knowledge base of these behaviors where as children do not. NFPA 4.5.1; 5.5.2. *CO, 2E,* Page 223.

Question #65. By utilizing engineering practices, fire suppression systems and building construction designs can control fire behavior and growth. NFPA 1021 5.5.1. *CO, 2E,* Page 224.

Question #66. The fire protection handbook published by the NFPA in 1997 (18th edition) puts the statistics together to back this. NFPA 1021 4.5.1; 5.5.2. *CO, 2E,* Page 221.

Question #67. Understanding the facility that you are in will help with safety. That is a key reason for pre-incident planning. NFPA 1021 - 5.5. *CO, 2E,* Pages 248, 266.

Question #68. Understanding the building construction will allow you to evaluate survival factors. NFPA 1021 - 5.5. *CO, 2E,* Page 252.

Question #69. Fire loads are more than the amount of materials: it is the quality and combustibility of the materials at which you must look. NFPA 1021 - 5.5. *CO, 2E,* Page 265.

Question #70. Understanding community demographics and the fire problem will allow you to assess the programs that are need and a true risk assessment to be done. NFPA 1021 - 5.5. *CO, 2E,* Page 265.

Question #71. Fire behavior is a key in determining the strategy and tactics during a working fire. NFPA 1021 - 4.6 AND 5.6. *CO, 2E,* Page 257.

Question #72. Fire behavior is a key in determining the strategy and tactics during a working fire. NFPA 1021 - 4.6 AND 5.6. *CO, 2E,* Page 266.

Question #73. Fire behavior is a key in determining the strategy and tactics during a working fire. NFPA 1021 - 4.6. *CO, 2E,* Page 262.

Question #74. Cause and determination must be a collective process of all individuals on the scene. Information of the initial crews and what they saw in fire behavior is critical in the information that an investigator must have to determine the cause and origin. NFPA 1021 - 4.5 AND 5.5.2. *CO, 2E,* Page 277.

Question #75. Legal proceedings are imperative to the conviction of arson. Understanding the legal processes and requirements is critical to the investigation process. NFPA 1021 - 4.5 AND 5.5.2. *CO, 2E,* Page 285.

Question #76. Legal proceedings are imperative to the conviction of arson. Understanding the legal processes and requirements is critical to the investigation process. NFPA 1021 - 4.5 AND 5.5.2. *CO, 2E,* Page 281.

Question #77. This cause of fire will be directly related to events. It is important to remember that lightning does not just occur during rain storms but under specific atmospheric conditions. NFPA 1021 - 4.5 AND 5.5.2. *CO, 2E,* Page 274.

Question #78. There are specific signs and particular events that must be present for this to be a just cause for a fire. NFPA 1021 - 4.5 AND 5.5.2. *CO, 2E,* Page 274.

Question #79. This is a natural progression caused by the fire. Any other action resulting in the same actions, caused by humans is called ventilation. NFPA 1021 - 4.5 AND 5.5.2. *CO, 2E,* Page 285.

Question #80. Legal proceedings are imperative to the conviction of arson. Understanding the legal processes and requirements is critical to the investigation process. NFPA 1021 - 4.5 AND 5.5.2. *CO, 2E,* Page 285.

Question #81. Careful recording of the observations made during the course of any fire and investigation may be used later especially if the case goes to trial. NFPA 1021 - 4.5 AND 5.5.2. *CO, 2E,* Page 283.

Question #82. The understanding of fire behavior is critical for any fire officer as they develop an action plan. NFPA 1021 - 4.5 AND 5.5.2. *CO, 2E,* Page 285.

Question #83. There are specific signs and particular events that must be present for this to be a just cause for a fire. NFPA 1021 - 4.5 AND 5.5.2. *CO, 2E,* Page 272.

Question #84. The majority of fire investigations will be from accidental causes and will be easy to determine. However, each investigation should be sequential and all aspects documented. NFPA 1021 - 4.5 AND 5.5.2. *CO, 2E,* Pages 272-274.

Question #85. Arson is a legal term. Under common law, arson was defined simply as the malicious burning of someone else's house. Today, the definition varies from state to state. Note your state's definition. NFPA 1021 - 4.5. *CO, 2E,* Pages 272-273.

Question #86. Fire control is known as both confining and extinguishing the fire. This means more than fire suppression. It includes overhaul as well to ensure all fire is found and suppressed. NFPA 1021 - 4.2.3, 4.6, 4.7.1 AND 5.6. *CO, 2E,* Page 314.

Question #87. You should conduct training using the pre-plan. NFPA 1021 - 4.6 AND 5.6. *CO, 2E,* Page 303.

Question #88. A plot plan allows for quick viewing of potential exposures and access points available to the fire department. This is key to operations. NFPA 1021 - 4.6 AND 5.6. *CO, 2E,* Page 314.

Question #89. We are concerned not only with the life safety of the civilians we protect, but also for the responders. NFPA 1021 - 4.6, 4.7.1, 5.7 AND 5.6. *CO, 2E,* Page 314.

Question #90. It is important that these activities be completed to have a successful incident response. NFPA 1021 - 4.6 AND 5.6. *CO, 2E,* Pages 290-291.

Question #91. This will never be a mode in which you start out. It is a time when you change from one strategy to another. NFPA 1021 - 4.2.1, 4.6 AND 5.6.2. *CO, 2E,* Page 351.

Question #92. Understanding the definition of an offensive mode is critical to fire officers as they make decisions as first-arrival units on scene that will be the deciding factors in how the incident runs and terminates. NFPA 1021 - 4.2.1, 4.6 AND 5.6.2. *CO, 2E,* Page 350.

Question #93. the various maneuvers that can be used to achieve a strategy while fighting fire or dealing with a similar emergency NFPA 1021 - 4.2.1, 4.6 AND 5.6.2. *CO, 2E,* Page 351.

Question #94. Understanding what mode to choose is critical to fire officers as they make decisions as first-arrival units on scene that will be the deciding factors in how the incident runs and terminates. NFPA 1021 - 4.2.1, 4.6 AND 5.6.2. *CO, 2E,* Pages 327-328.

Question #95. This is a message that the fire is under control and a critical piece of information for the incident commander as they work through the incident priorities. NFPA 1021 - 4.2.1, 4.6.4 AND 5.6.2. *CO, 2E,* Page 321.

Question #96. It is important that company officers know weather and other environmental conditions. NFPA 1021 - 4.2.1, 4.6 AND 5.6.2. *CO, 2E,* Page 326.

Question #97. This is chosen as a mode to limit the amount of risk posed to the firefighter. NFPA 1021 - 4.2.1, 4.6 AND 5.6.2. *CO, 2E,* Pages 327-328.

Question #98. NFPA standards set necessary proficiencies that are required in certain situations and required for proficiency by firefighters. NFPA 1021 - 4.6, 4.6.3, 5.6.1 AND 5.6.2. *CO, 2E,* Page 343.

Question #99. To run a large operation or one of unusual nature, it is important to be able to get the resources necessary to mitigate the incident. NFPA 1021 - 4.2.1, 4.6, 5.6.1 AND 5.6.2. *CO, 2E,* Page 347.

Question #100. These are crucial pieces to the overall strategic planning process of any incident. It allows you to evaluate progress and tactics over a set period of time and to maintain focus on those. NFPA 1021 - 4.2.1, 4.6 AND 5.6.2. *CO, 2E,* Page 350.

PHASE I

Exam III

1. According to the text, the word ______ means that an individual has been tested by an accredited examination body on clearly identified material and found to meet a minimum standard.
   a. licensing
   b. certification
   c. qualified
   d. bonded

2. The capabilities, efficiency, and morale of the company are direct reflections of the company officer's _________.
   a. time management
   b. seniority
   c. leadership ability
   d. general knowledge

3. We can prepare to make the transition from firefighter to officer with __________.
   a. study and effort
   b. practice and patience
   c. hard work
   d. physical abilities

4. A shortcoming of too many officers is the lack of __________.
   a. experience
   b. education
   c. communication skills
   d. training

5. In fire departments or other emergency response organizations, what percent of the employees work at the company level?
   a. 80% to 90%
   b. 60% to 75%
   c. 50% to 65%
   d. 30% to 45%

6. Being a role model means being a professional. Professionalism encompasses which of the following?
   a. being a coach, evaluator, leader, and a supervisor
   b. attitude, behavior, communication, demeanor, and ethics
   c. being a decision-maker, friend, planner, and a supervisor
   d. being a manager, motivator, referee, and an innovator

7. The term company is used to describe __________.
   a. the department
   b. work teams
   c. strike teams
   d. a senior officer
8. The following principles are part of effective writing in your communication skills except ______.
   a. consider the reader
   b. technical terms
   c. simplicity
   d. emphasis
9. Which of the following is a barrier to effective communications?
   a. physical barrier
   b. personal barrier
   c. semantic barrier
   d. all of the above
10. Walls, distance, and background noise are examples of __________.
    a. physical barriers to effective communications
    b. things that provide for effective communications
    c. personal barriers to effective communications
    d. semantic barriers to effective communications
11. All of the following are parts of the communication model except __________.
    a. sender
    b. medium
    c. message
    d. response
12. When you actively listen, __________________.
    a. you actually hear what is said
    b. you show respect for the sender
    c. you will remember what is said
    d. all of the above
13. The organizational principle whereby there is only one supervisor refers to ____________.
    a. unity of command
    b. chain of command
    c. scalar principle
    d. division of labor

14. An important empowerment tool that generally makes the subordinates feel better about their jobs is __________.
    a. delegation
    b. education
    c. leadership
    d. communications

15. Firefighters informally evaluate their company officers on ___________.
    a. how the officers perform their duties
    b. how well the officers apply policy
    c. how fair the officers are among members
    d. all of the above

16. The definition of ________ is being responsible for one's personal activities; in the organizational context, it includes being responsible for the actions of one's subordinates.
    a. responsibility
    b. accountability
    c. line authority
    d. management

17. The definition of ________ is being accountable for the actions and activities, and having moral and perhaps legal obligation to carry out certain activities.
    a. responsibility
    b. accountability
    c. line authority
    d. management

18. Most fire departments in America consist of ________ companies.
    a. paid
    b. volunteer
    c. social
    d. dedicated

19. The organizational principle whereby there is a continuous chain of command refers to ____________.
    a. unity of command
    b. chain of command
    c. scalar principle
    d. division of labor

20. The success of any organization is dependent upon _________.
    a. funding and structure
    b. leaders and followers
    c. structure and leaders
    d. strong leadership and strong communications

21. If a company officer assigns duties to their subordinates, they are _________.
    a. negotiating
    b. commanding
    c. delegating
    d. ordering
22. Activities that support those providing emergency services are said to be ________.
    a. support staff
    b. secretaries
    c. staff functions
    d. assignments
23. Span of control is ___________.
    a. delegation
    b. micro-management
    c. the organizational principle of one supervisor
    d. the number of people that can be effectively managed
24. ________ exist whenever two or more people share a common goal.
    a. Teams
    b. Units
    c. Groups
    d. Councils
25. The first step in the management process is _________.
    a. organizing
    b. coordinating
    c. planning
    d. commanding
26. A system of conduct, principles of honor and morality, or guidelines for human actions describe ________.
    a. loyalty
    b. human rights
    c. ethics
    d. management
27. Which describes the time on which you focus in long-range planning?
    a. one to five years
    b. within one year
    c. beyond five years
    d. none of the above

28. ____________ is monitoring our progress.
    a. Organizing
    b. Commanding
    c. Coordinating
    d. Controlling

29. Management can be enhanced through good _____, ________, and even by personal observation.
    a. leadership, motivation
    b. communication, structure
    c. policy, procedures
    d. leadership, policies

30. ____________ is the right and power to command.
    a. Control
    b. Intimidation
    c. Authority
    d. Responsibility

31. ____________ involves a third party acting as a facilitator.
    a. Mediation
    b. Arbitration
    c. A union
    d. Negotiation

32. The act of seeking advice or gathering information from another is known as ________.
    a. influencing
    b. consulting
    c. directing
    d. supporting

33. A recognition of authority derived from the government or other appointing agency best describes ___________ power.
    a. reward
    b. expert
    c. legitimate
    d. organizational

34. The first Civil Rights Act was passed by Congress in ________.
    a. 1954
    b. 1972
    c. 1963
    d. 1964

35. The power bestowed upon an individual by the organization is referred to as ________ power.
    a. licensed
    b. legitimate
    c. legal
    d. legislative

36. The personal actions needed by managers to get employees to carry out certain activities define ____________.
    a. oral counseling
    b. authority
    c. leadership
    d. responsibility

37. The first expression of dissatisfaction might be considered a _______________.
    a. conflict
    b. grievance
    c. complaint
    d. gripe

38. Three effective tools for employee development are ___________.
    a. education, opportunity, and advancement
    b. coaching, counseling, and evaluations
    c. education, counseling, and evaluations
    d. coaching, opportunities, and evaluations

39. The first step in a formal disciplinary process is ________________.
    a. investigating statements
    b. fact gathering
    c. oral reprimand
    d. oral counseling

40. ____________ is an administrative process whereby an employee is punished for not conforming to the organizational rules and regulations.
    a. Disciplinary action
    b. Arbitration action
    c. Termination action
    d. Written reprimand

41. The step in the disciplinary process that provides the employee a fresh start in another venue would be known as a ________.
    a. suspension
    b. transfer
    c. demotion
    d. termination

42. __________ is an informational process that helps subordinates improve their skills and abilities.
    a. Counseling
    b. Coaching
    c. Evaluation
    d. Meeting

43. When evaluating a team member, the term ________ means to over weight certain factors among the overall information available.
    a. "like-me" syndrome
    b. contamination
    c. halo effect
    d. central tendency

44. Repeated dissatisfaction over the topic might lead to a _________.
    a. complaint
    b. gripe
    c. grievance
    d. conflict

45. Someone who helps another develop a skill is called a(n) _____________.
    a. counselor
    b. coach
    c. evaluator
    d. none of the above

46. Getting employees involved is called _____________.
    a. empowerment
    b. arbitration
    c. coaching
    d. evaluation

47. The standard on Fire Department Occupational Safety and Health program is NFPA ______.
    a. 1021
    b. 2100
    c. 1910
    d. 1500

48. Which one of the following is not part of the five generally accepted tools for dealing with critical incident stress?
    a. training
    b. peer support
    c. counseling
    d. vacation

49. The largest percentage of firefighter injuries and deaths occur while __________.
    a. at the training exercises
    b. responding to calls
    c. on the fire ground
    d. at the fire station

50. The standard on Fire Department Occupational Safety and Health program for the Health and Safety Officer is NFPA ______.
    a. 1021
    b. 2100
    c. 1910
    d. 1500

51. The tracking of personnel as to location and activity during an emergency event is known as __________.
    a. rapid intervention teams
    b. personnel accountability
    c. emergency operations
    d. company officer

52. Action taken to prevent a fire from occurring or, if one does occur, to minimize the loss is called _________.
    a. fire control
    b. fire prevention
    c. fire inspection
    d. fire suppression

53. _________ was published in 1973 and was a report of a study by the National Commission of Fire Prevention and Control.
    a. American Burning
    b. American Fire
    c. America Burning
    d. America Fire

54. _______ is the leading cause of fires in residences and is by far the leading cause of fire-related injuries in residences.
    a. Arson
    b. Cooking
    c. Electrical
    d. A child

55. A fire prevention tool required where there is potential for life loss, or where there are hazardous materials or hazardous processes is called a ______.
    a. permit
    b. certificate
    c. license
    d. standard

56. A document serving as evidence of the completion of an educational or training program, or a document issued to an individual or company as a fire prevention tool is a __________.
    a. permit
    b. certificate
    c. license
    d. standard

57. Law or regulation that establishes minimum requirements for the design and construction of buildings are the __________ codes.
    a. fire prevention
    b. standard
    c. NFPA fire
    d. building

58. Action taken to control and extinguish a fire is called __________.
    a. fire control
    b. fire prevention
    c. fire inspection
    d. fire suppression

59. A document serving as evidence of the completion of an educational or training program, or a document issued to an individual or company as a fire prevention tool is a __________.
    a. permit
    b. certificate
    c. license
    d. standard

60. Legal documents that set forth the requirements for life safety and property protection in the event of fire, explosion, or similar emergency are __________ codes.
    a. fire prevention
    b. standard
    c. NFPA
    d. building

61. Type ____ construction is called "woodframe," which is completely vulnerable to fire.
    a. II
    b. III
    c. IV
    d. V

62. As defined, _____________ is an assessment of the consequences on the community, which includes the people, their property, and the environment.
    a. access factors
    b. physical factors
    c. occupancy factors
    d. community consequences

63. What are the three stages of fire growth within an enclosed structure?
    a. incipient phase, thermal phase, and smoldering phase
    b. free-burning phase, thermal phase, and smoldering phase
    c. incipient phase, free-burning phase, and smoldering phase
    d. smoldering phase, incipient phase, and thermal phase

64. The ________________ is the amount of heat required to convert a substance from a liquid to a vapor.
    a. specific heat
    b. British thermal unit
    c. latent heat of vaporization
    d. heat of combustion

65. Ventilation is accomplished in two ways, generally referred to as _______ ventilation and ________ventilation.
    a. positive, negative
    b. natural, mechanical
    c. open, closed
    d. none of the above

66. The __________________ is the amount of heat required to raise the temperature of one pound of water one degree Fahrenheit.
    a. specific heat
    b. British thermal unit
    c. latent heat of vaporization
    d. heat of combustion

67. What describes the spread of the fire through the fuel load and the structure itself?
    a. fire extinguishment
    b. fire extension
    c. backdraft
    d. flashover

68. Fire scenes are documented by all of the following except _________________.
    a. notes
    b. diagrams
    c. photographs
    d. blueprints

69. Arson fires are referred to as __________.
    a. incendiary
    b. suspicious
    c. intentionally set
    d. all of the above

70. Common natural causes of fires are ______________ .
    a. cooking
    b. electricity
    c. smoking
    d. none of the above

71. Common arson causes of fires are ______________ .
    a. cooking
    b. electricity
    c. smoking
    d. heating equipment
    e. none of the above

72. Common natural causes of fires are ______________ .
    a. lightning
    b. electricity
    c. smoking
    d. all of the above

73. According to the text, a majority of fires are ____________________ in nature.
    a. suspicious
    b. accidental
    c. arson
    d. intentional

74. Common accidental causes of fire include all of the following except __________.
    a. cooking
    b. heating
    c. smoking
    d. arson

75. ___________ is defined as opened to the atmosphere by the fire burning through windows or walls through which heat and fire by-products are released, and through which fresh air may enter.
    a. Oxidation
    b. Arson
    c. Vented
    d. Ventilation

76. Three methods by which the fire department may reenter a property after the fire is extinguished include all of the following except __________.
    a. owner's consent or consent to search
    b. administrative search warrant
    c. investigative warrant
    d. search warrant

77. Arson is the ______________________ .
    a. burning of dwellings
    b. burning of buildings other than dwellings
    c. burning of other property
    d. attempted burning of buildings or property
    e. all of the above

78. Common natural causes of fires are ______________ .
    a. autoignition through chemical reaction
    b. electricity
    c. lightning
    d. answer (a) and (c)
    e. answer (b) and (c)

79. What is defined as the first priority during all emergency operations and addresses the safety of occupants and emergency responders?
    a. life safety
    b. fire control
    c. property conservation
    d. area of refuge

80. A __________ is a bird's-eye view of the structure with the roof removed showing walls, doors, and stairs.
    a. plot plan
    b. floor plan
    c. blueprint
    d. format layout

81. The third priority during all emergency operations that addresses the safety of occupants and emergency responders is __________ .
    a. fire control
    b. life safety
    c. property conservation
    d. access

82. ___________ is an activity in comprehensive emergency management.
    a. Mitigation
    b. Recovery
    c. Response
    d. Preparedness
    e. All of the above

83. The first step in a pre-incident survey process is to _________________________ .
    a. make a list of fire hazards
    b. make an appointment
    c. make a plot plan
    d. all of the above

84. What is defined as the effort to reduce primary and secondary damage?
    a. life safety
    b. fire control
    c. property conservation
    d. area of refuge

85. What is defined as the first priority during all emergency operations that addresses the safety of occupants and emergency responders?
    a. life safety
    b. fire control
    c. property conservation
    d. area of refuge

86. Target hazard facilities include ________________________ .
    a. nursing homes
    b. small storage buildings
    c. residential houses
    d. all of the above

87. What sets broad goals and outlines the overall plan to control the incident?
    a. action plan
    b. strategy
    c. tactics
    d. none of the above

88. What is the definition of offensive mode?

    a. firefighting operations that make a direct attack on a fire for purposes of control and extinguishment
    b. actions intended to control a fire by limiting its spread to a defined area
    c. the critical process of shifting from the offensive to the defensive mode or from the defensive to offensive
    d. none of the above

89. As defined, a task is ____________________ .

    a. the firefighting operations that make a direct attack on a fire for purposes of control and extinguishment
    b. the duties and activities performed by individuals, companies, or teams
    c. a broad set of goals and outlines of the overall plan to control the fire
    d. the various maneuvers that can be used to achieve a strategy while fighting a fire or dealing with a similar emergency
    e. none of the above

90. The second priority of the fire officer at the scene of an emergency is __________.

    a. life safety
    b. property conservation
    c. incident stability
    d. benchmarks

91. ______________ are significant points in the emergency event usually marking the accomplishment of one of the three incident priorities.

    a. Benchmarks
    b. Stopping points
    c. Signals
    d. Command stops

92. The ____________________ section is responsible for collecting, evaluating, and disseminating appropriate information.

    a. operations
    b. planning
    c. logistics
    d. finance/administration
    e. none of the above

93. Incident priority benchmarks are _____________________ .

    a. incident stabilization
    b. life safety
    c. property conservation
    d. only (a) and (b)
    e. answer (a), (b), and (c)

94. When transmitting "benchmarks" to the command for property conservation, the terminology used should be ___________ .
    a. "all clear"
    b. "under control"
    c. "loss stopped"
    d. "all OK"

95. The third priority of the fire officer at the scene of an emergency is __________.
    a. life safety
    b. property conservation
    c. incident stability
    d. benchmarks

96. ______________ are significant points in the emergency event usually marking the accomplishment of one of the three incident priorities.
    a. Benchmarks
    b. Stopping points
    c. Signals
    d. Command stops

97. ______________ is used to provide actions intended to control a fire by limiting its spread to a defined area.
    a. Offensive attack
    b. Defensive mode
    c. Transitional mode
    d. Offensive mode

98. Sectors, sections, divisions, and groups belong under the ____________________ section.
    a. operations
    b. planning
    c. logistics
    d. finance/administration
    e. none of the above

99. The Demobilization Unit belongs under the ____________________ section.
    a. operations
    b. planning
    c. logistics
    d. finance/administration
    e. none of the above

100. The Incident Action Plan is _______________ .

   a. a vivid but brief description of the on-scene conditions relevant to the emergency

   b. a list of assistance to be provided by another fire department or agency

   c. an organized course of action that addresses all phases of incident command within a specific period of time

   d. none of the above

Phase I, Exam III: Answers to Questions

1. B
2. C
3. A
4. C
5. A
6. B
7. B
8. B
9. D
10. A
11. D
12. D
13. A
14. A
15. D
16. B
17. A
18. B
19. C
20. D
21. C
22. C
23. D
24. C
25. C
26. C
27. C
28. D
29. C
30. C
31. A
32. B
33. C
34. D
35. B
36. C
37. D
38. B
39. C
40. A
41. B
42. B
43. C
44. A
45. B
46. A
47. D
48. D
49. C
50. D
51. B
52. B
53. C
54. B
55. A
56. B
57. D
58. D
59. B
60. A
61. D
62. D
63. C
64. C
65. B
66. B
67. B
68. D
69. D
70. D
71. E
72. A
73. B
74. D
75. C
76. C
77. E
78. D
79. A
80. B
81. C
82. E
83. B
84. C
85. A
86. A
87. B
88. A
89. B
90. C
91. A
92. B
93. E
94. C
95. B
96. A
97. B
98. A
99. B
100. C

Phase I, Exam III:
Rationale & References for Questions

Question #1. Certification provides a yard stick for measuring competency in every type of fire department. NFPA 1021 4.1.1. *CO, 2E,* Page 8.

Question #2. Leadership is the company officer's primary job. The success of the company is directly proportional to the officer's leadership ability. NFPA 1021 4.2-4.2.6. *CO, 2E,* Page 5.

Question #3. There is much to learn to become a company officer. Preparation is the key to being successful in this quest. NFPA 1021 4.1.1. *CO, 2E,* Page 12.

Question #4. Communication skills are important in every aspect of our lives including both personal and professional. Communication skills are an important part of the everyday life of a company officer. NFPA 4.1.1. *CO, 2E,* Page 14.

Question #5. If you examine any organizational chart within a department, you will find the majority of the work force located at the company level. These individuals are the ones who work to carry out the mission of the department. NFPA 1021 4.2 -4.7; 5.2-5.7.1. *CO, 2E,* Page 7.

Question #6. Professionalism is a term used by individuals to describe their particular characteristics. These characteristics can be summed up as it is as easy as A B C D E. NFPA 1021 4.2 -4.7. *CO, 2E,* Pages 14-15.

Question #7. Companies are the basic group of work force in the fire service. 80 - 90 % of the work done is accomplished at the company level. NFPA 1021 4.2.1; 4.2.2. *CO, 2E,* Page 5.

Question #8. Technical terms may be understood by yourself or other fire service professionals; however, they are not understood by the general public. It is important to use the simplistic approach when writing to guarantee that effective communications is occurring. NFPA 1021 4.2.1; 5.2.1. *CO, 2E,* Page 28.

Question #9. Consider a barrier to be like a filter. The more barriers you have, the less that goes through. Thus, the less communication or efficiency that you have in communications. NFPA 1021 4.2.5; 5.4.3. *CO, 2E,* Pages 23-24.

Question #10. Consider a barrier to be like a filter. The more barriers you have, the less that goes through. Thus, the less communication or efficiency that you have in communications. NFPA 1021 4.2.1-4.3; 5.2.1- 5.4.5. *CO, 2E,* Page 23.

Question #11. Regardless of whether you are speaking or writing, the communications process is generally thought to include five elements, called the communications model. NFPA 1021 4.2.5; 5.4.3. *CO, 2E,* Page 21.

Question #12. Understanding others requires an active role on the part of the listener. NFPA 1021 4.2. *CO, 2E,* Page 25.

Question #13. This concept is frequently confused with chain of command. Unity of command is an essential organizational concept. NNFPA 1021 4.1.1; 5.1.1. *CO, 2E,* Page 66.

Question #14. Delegation shows trust and confidence in the group or firefighter. It is important to use this power with caution so as not to delegate a task to an individual that is not capable of handling it. This will cause ineffectiveness in the process and will cause the firefighter to be set up for failure. NFPA 1021 4.2; 5.6. *CO, 2E,* Page 62.

Question #15. This is a naturally occurring process as subordinates will look at the aspects of how well the company officer is doing their job. This also makes the company officer perform at a higher level by striving to set a good example and be a strong leader. NFPA 4.2; 5.2. *CO, 2E,* Page 65.

Question #16. The company officer position is one that requires the individual filling that role to now be responsible for their actions and the actions of others. This is a large step in the organization chain. NFPA 1021 4.4; 5.4.5. *CO, 2E,* Page 62.

Question #17. As a company officer you will have set policy and procedures that must be followed. You have a responsibility to follow rules, regulations, set laws, and ethics will pick up where laws leave off. NFPA 1021 4.4; 5.4.5. *CO, 2E,* Page 62.

Question #18. The majority of the United States is rural and the majority of the population in these areas are protected by volunteer firefighters. As the size of the population in an area increases, so does the demands for service based upon the risks and target hazards that follow. NFPA 1021 4.1.1; 5.1.1. *CO, 2E,* Page 54.

Question #19. Like playing a scale on a musical instrument where every note is sounded, the scaler principle suggests that every level in the organization is considered in the flow of communications. NNFPA 1021 4.1.1; 5.1.1. *CO, 2E,* Page 66.

Question #20. Most organizations have a pyramid structure, with one person in charge, and an increasing number of subordinates at each level as you move downward in the organization. These levels require strong leadership whereas the majority of the subordinates are at the company level. NFPA 1021 4.1.1; 5.1.1. *CO, 2E,* Page 60.

Question #21. Delegating allows for an increased amount of work to be done, thus accomplishing more goals. NFPA 1021 4.2; 5.6. *CO, 2E,* Page 61.

Question #22. Staff functions do not normally get directly involved with delivering emergency services. NFPA 1021 4.4; 5.4. *CO, 2E,* Page 64.

Question #23. It is important to realize that the span of control is between four and seven. Depending on the efficiency and leadership abilities of the company officer will depend on the amount of personnel they will be able to lead effectively. NFPA 1021 4.2; 5.6. *CO, 2E,* Page 67.

Question #24. Organizations are groups of people. NFPA 1021 4.1.1; 5.1.1. *CO, 2E,* Page 44.

Question #25. Planning is largely a mental process and often requires the ability to envision things that have not yet happened. NFPA 1021 4.2; 5.4. *CO, 2E,* Page 77.

Question #26. Ethics have a direct impact on the management of the fire service, from chief to company officcr. NFPA 1021 4.1.1. *CO, 2E,* Page 89.

Question #27. It is important to do long-range planning to try to keep ahead of the community's needs. NFPA 1021 4.6.2; 5.4.2. *CO, 2E,* Page 77.

Question #28. If planning is looking ahead to see what we are going to have to do, controlling is looking back to see how well we have done. NFPA 1021 4.2.6. *CO, 2E,* Page 78.

Question #29. Policies and procedures give us guidelines to follow as we manage personnel based upon the organization's leadership. NFPA 1021 4.2; 5.2.1. *CO, 2E,* Page 84.

Question #30. To be effective, delegation of the duty and acceptance of the obligation for the job should be coupled with the authority to do the job. NFPA 1021 4.2; 5.2. *CO, 2E,* Page 99.

Question #31. The mediator opens communication channels. NFPA 1021 4.1.1; 5.1.1. *CO, 2E,* Page 102.

Question #32. Here there is discussion in which the supervisor seeks ideas, explains the needs and decisions, and sells the idea. NFPA 1021 4.2.2; 5.2. *CO, 2E,* Page 128.

Question #33. With legitimate power, two other types of power are implied: reward and punishment. As a company officer, once promoted, you automatically receive these powers. NFPA 1021 4.1.1; 5.2. *CO, 2E,* Page 127.

Question #34. This was the act that authorized equal employment opportunity. NFPA 1021 4.1.1; 5.1.1. *CO, 2E,* Page 130.

Question #35. With legitimate power, two other types of power are implied: reward and punishment. As a company officer, once promoted, you automatically receive these powers. NFPA 1021 4.1.1; 5.2. *CO, 2E,* Page 127.

Question #36. A leader is a person who has the ability to get other people to do what they don't want to do, and like it. NFPA 1021 4.1.1; 5.2.1. *CO, 2E,* Page 120.

Question #37. Gripes should be channeled into positive action. NFPA 1021 4.2.4; 5.2.1. *CO, 2E,* Page 163.

Question #38. There are four effective tools for team member development. Although the role is critical to the collective productivity and safety of the members assigned to that company, we should also look at the role of the supervisor as it relates to each member as an individual. NFPA 1021 4.2; 5.2.2. *CO, 2E,* Page 142-154.

Question #39. This is much like a private counseling session between the supervisor and the member. NFPA 1021 4.2.4; 5.2.1. *CO, 2E,* Page 161.

Question #40. When taking a disciplinary action, think of it as another way of improving member performance and behavior. NFPA 1021 4.2.4; 5.2.1. *CO, 2E,* Page 160.

Question #41. This may provide a solution to the problem. Be careful not to shift the problem to someone else. This works well in larger departments, but not as well in smaller ones. NFPA 1021 4.2.4; 5.2.1. *CO, 2E,* Page 161.

Question #42. Coaching usually focuses on one aspect of the job at a time. NFPA 1021 4.2.4; 5.7. *CO, 2E,* Page 147.

Question #43. This concept results in a score that is distorted either too high or too low. NFPA 1021 4.2.4; 5.2.1. *CO, 2E,* Page 155.

Question #44. Complaints should be channeled into a positive action. NFPA 1021 4.2.4; 5.2.1. *CO, 2E,* Page 163.

Question #45. Coaching is an informational process that helps members improve their skills and abilities. NFPA 1021 4.2.4; 5.7. *CO, 2E,* Page 147.

Question #46. Empowerment allows team members to have a feeling of ownership in the organization. NFPA 4.1.1; 5.2.1. *CO, 2E,* Page 162.

Question #47. NFPA 1500 was designed to help fire departments reduce the frequency and severity of accidents and injuries to its members. NFPA 4.1.1; 5.1.1. *CO, 2E,* Page 183.

Question #48. Critical incident stress cannot be avoided, but its effects can be lessened by having a support system in place. NFPA 1021 4.2.4; 5.2.1. *CO, 2E,* Page 206.

Question #49. These are statistics from 2002 as published by the National Fire Protection Association. NFPA 1021 4.7; 5.7. *CO, 2E,* Page 179.

Question #50. Every department should have an individual assigned to the duties of health and safety officer. NFPA 4.1.1; 5.1.1. *CO, 2E,* Page 183.

Question #51. Being able to keep track of firefighters and their location is important for the true accounting of the personnel on scene and will help the Rapid Intervention Team if they are required to help locate a lost or missing firefighter. NFPA 4.2.1; 5.6.1. *CO, 2E,* Page 197.

Question #52. This is a proactive approach that will help lessen the devastating effects of fires and other disasters. NFPA 1021 4.6.1; 5.5. *CO, 2E,* Page 214.

Question #53. This report noted that fire prevention activities were frequently assigned a low priority in fire departments. It highlights the problems identified by the commission such as a need for more emphasis on fire prevention, a need for better training for the fire service, and a need for more fire safety education for the public. NFPA 1021 4.1.1. *CO, 2E,* Page 214.

Question #54. Many of these fires are the result of unattended cooking. NFPA 1021 4.5.1; 5.5.2. *CO, 2E,* Page 221.

Question #55. Permits authorize a specific activity, provide conditions that must be satisfied for that activity to take place, and provide notice to the nearest fire company and other governmental agencies. NFPA 1021 5.5.1. *CO, 2E,* Page 227.

Question #56. These certificates allow for a measurable outcome of performance. NFPA 1021 4.3; 5.5. *CO, 2E,* Page 227.

Question #57. These codes provide an enforcement authority. NFPA 1021 4.1.1; 5.5. *CO, 2E,* Page 228.

Question #58. Fire suppression is putting out all fire. This is when resources are activated after the incident has occurred. NFPA 1021 4.2.1. *CO, 2E,* Page 214.

Question #59. Licenses allow for a set measured level of expertise proven generally through education and testing. NFPA 1021 4.3; 5.5. *CO, 2E,* Page 227.

Question #60. These codes provide enforcement authority. NFPA 1021 4.1.1; 5.5. *CO, 2E,* Page 228.

Question #61. Woodframe construction has always presented a fire problem since the pioneer days. NFPA 1021 - 5.5. *CO, 2E,* Page 252.

Question #62. Understanding the community's fire problem is critical in the planning process along with the development of training and other aspects of the company officer's routines. NFPA 1021 - 5.5. *CO, 2E,* Page 265.

Question #63. A true understanding of fire behavior is critical in the development of action plans upon arrival to fire scenes. NFPA 1021 - 4.6 AND 5.6. *CO, 2E,* Page 257.

Question #64. In fire suppression activities, the most common extinguishing agent is water. Understanding the concept of latent heat of vaporization is advantageous since water is converted to steam and can absorb more Btus, thus extinguishing the fire by taking away one side of the fire tetrahedron. NFPA 1021 - 4.6 AND 5.6. *CO, 2E,* Page 262.

Question #65. Ventilation can be described as the procedures necessary to effect the planned and systematic direction and removal of smoke heat and fire gases from within a structure. NFPA 1021 - 4.6 AND 5.6. *CO, 2E,* Page 261.

Question #66. Every pound of water will absorb 970 additional BTUs. NFPA 1021 - 4.6. *CO, 2E,* Page 265.

Question #67. The building construction will dictate the amount of fire extension you will have. Woodframe, especially balloon-frame construction, will allow for more fire extension that platform woodframe construction. Fire behavior and building construction are related in that fire will only behave in the building as the building construction will allow the fire to behave. NFPA 1021 - 4.6. *CO, 2E,* Page 261.

Question #68. These are key especially in a fire investigation. NFPA 1021 - 4.5 AND 5.5.2. *CO, 2E,* Page 283.

Question #69. Arson is a legal term. Arson is a deliberately set fire, but not all intentionally set fires are arson. Arson has two elements: a criminal intent and deliberate or intentional act. NFPA 1021 - 4.5 AND 5.5.2. *CO, 2E,* Page 271.

Question #70. All are accidental causes. But the only natural cause would be electricity or spontaneous combustion. NFPA 1021 - 4.5 AND 5.5.2. *CO, 2E,* Page 274.

Question #71. All are accidental causes. Arson is deliberately set fires. NFPA 1021 - 4.5 AND 5.5.2. *CO, 2E,* Page 274.

Question #72. Natural causes of fires are ones that are a direct result of nature and science. NFPA 1021 - 4.5 AND 5.5.2. *CO, 2E,* Page 274.

Question #73. Human factors are generally involved in accidental fires. NFPA 1021 - 4.5 AND 5.5.2. *CO, 2E,* Page 271.

Question #74. Arson is a direct result of a deliberately set fire. NFPA 1021 - 4.5. *CO, 2E,* Pages 272-273.

Question #75. The natural occurring of an opening occurring during the burning process is called vented. When we open the structure for a planned process of the removal of smoke and fire products, we call this ventilation. NFPA 1021 - 4.5 AND 5.5.2. *CO, 2E,* Page 285.

Question #76. It is important to understand the legal process to help in the conviction of criminal acts as related to arson fires. NFPA 1021 - 4.5 AND 5.5.2. *CO, 2E,* Page 281.

Question #77. Arson is a legal term. Under common law, arson was defined simply as the malicious burning of someone else's house. This varies today from state to state, but generally the definition has been extended to include any property including one's own and designated four levels or degrees of arson. NFPA 1021 - 4.5 AND 5.5.2. *CO, 2E,* Page 274.

Question #78. Natural causes of fires are ones that are a direct result of nature and science. NFPA 1021 - 4.5 AND 5.5.2. *CO, 2E,* Page 274.

Question #79. As responders and especially as a company officer, your first priority on any scene is the life safety one; first the safety of your personnel and then the safety of civilians. NFPA 1021 - 4.6, 4.7.1, 5.5.1 AND 5.6. *CO, 2E,* Page 314.

Question #80. This plan will allow you to have a quick view of potential exposures along with quick access points. NFPA 1021 - 4.6 AND 5.6. *CO, 2E,* Page 314.

Question #81. To preserve as much property as possible is always a goal. However, life safety and incident stabilization come prior to property conservation. NFPA 1021 - 4.6, 4.7.1, 5.7 AND 5.6. *CO, 2E,* Page 314.

Question #82. These activities are common to all types of events, ranging from weather-related disasters to terrorist attacks. NFPA 1021 - 4.6 AND 5.6. *CO, 2E,* Pages 290-291.

Question #83. It is important to be able to have someone familiar with the building present to be able to allow you access to all areas and answer any questions you may have. NFPA 1021 - 4.6 AND 5.6. *CO, 2E,* Page 294.

Question #84. This is the third priority in any incident. NFPA 1021 - 4.6, 5.5.1 AND 5.6. *CO, 2E,* Page 314.

Question #85. As responders and especially as a company officer, your first priority on any scene is the life safety one; first the safety of your personnel and then the safety of civilians. NFPA 1021 - 4.6, 4.7.1, 5.5.1 AND 5.6. *CO, 2E,* Page 314.

Question #86. B. and C. are not high hazard structures. NFPA 1021 - 4.6 AND 5.6. *CO, 2E,* Page 293.

Question #87. Strategy is fundamental to planning and directing effective operations. NFPA 1021 - 4.2.1, 4.6 AND 5.6.2. *CO, 2E,* Page 351.

Question #88. This is a mode in which inherent risks are taken to control the fire and extinguish it. Most often, this operational mode is done interiorly. NFPA 1021 - 4.2.1, 4.6 AND 5.6.2. *CO, 2E,* Page 350.

Question #89. As a company officer, generally you delegate tasks to firefighters to get more work accomplished. NFPA 1021 - 4.2.1, 4.6 AND 5.6.2. *CO, 2E,* Page 351.

Question #90. You cannot be accountable for what happened prior to your arrival, but you can be held accountable for what happens after you arrive. You must be sure that your information, your planning, and your actions all support an effective, coordinated, and safe operation. NFPA 1021 - 4.2.1, 4.6 AND 5.6.2. *CO, 2E,* Page 320.

Question #91. Benchmarks allow the incident commander to keep up with critical points that will help measure the progression of the incident. NFPA 1021 - 4.2.1, 4.6 AND 5.6.2. *CO, 2E,* Page 350.

Question #92. The planning sector keeps track of the current situation, predicts the probable course of events, and prepares optional strategies and tactics. NFPA 1021 - 4.2.1, 4.6, 5.6.1 AND 5.6.2. *CO, 2E,* Page 347.

Question #93. Benchmarks allow the incident commander to keep up with critical points that will help measure the progression of the incident. NFPA 1021 - 4.2.1, 4.6 AND 5.6.2. *CO, 2E,* Pages 320-321.

Question #94. Benchmarks allow the incident commander to keep up with critical points that will help measure the progression of the incident. NFPA 1021 - 4.2.1, 4.6.4 AND 5.6.2. *CO, 2E,* Page 321.

Question #95. Property conservation is important and it will help with your quest for good customer service. NFPA 1021 - 4.2.1, 4.6 AND 5.6.2. *CO, 2E,* Page 320.

Question #96. Benchmarks allow the incident commander to keep up with critical points that will help measure the progression of the incident. NFPA 1021 - 4.2.1, 4.6 AND 5.6.2. *CO, 2E,* Page 350.

Question #97. During this mode, operations are conducted from a safe distance outside of the structure and are focused at containing the fire to a specific area rather than extinguishing it. NFPA 1021 - 4.2.1, 4.6 AND 5.6.2. *CO, 2E,* Page 350.

Question #98. Operations is responsible for management of all operational activity associated with the primary mission of IMS. NFPA 1021 - 4.2.1, 4.6, 5.6.1 AND 5.6.2. *CO, 2E,* Page 346.

Question #99. The planning sector keeps track of the current situation, predicts the probable course of events, and prepares optional strategies and tactics. NFPA 1021 - 4.2.1, 4.6, 5.6.1 AND 5.6.2. *CO, 2E,* Page 346.

Question #100. All operations should be conducted in an organized manner and an action plan will help you do just that. Remembering that we cannot control what has happened prior to our arrival and knowing we can be held responsible for what happens after our arrival, it is important to have the correct information and approach to control or mitigate the incident. NFPA 1021 - 4.2.1, 4.6 AND 5.6.2. *CO, 2E,* Page 350.

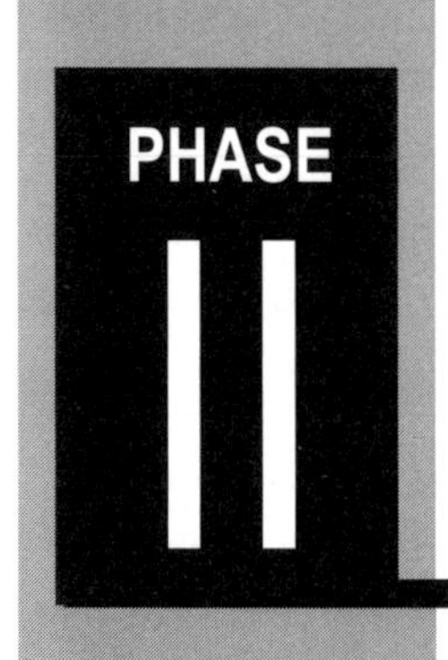

APPLICATION & ANALYSIS

Section two is evaluating for a higher level of learning. Within this section, we are testing to determine an understanding of comparing material, describing processes, explaining procedures, and interpreting results. A test-taker mastering this section should have a better grasp of the material and a greater depth of understanding. Referring to Table I-1 (Bloom's Taxonomy, Cognitive Domain), we are addressing the following levels:

- application
- analysis

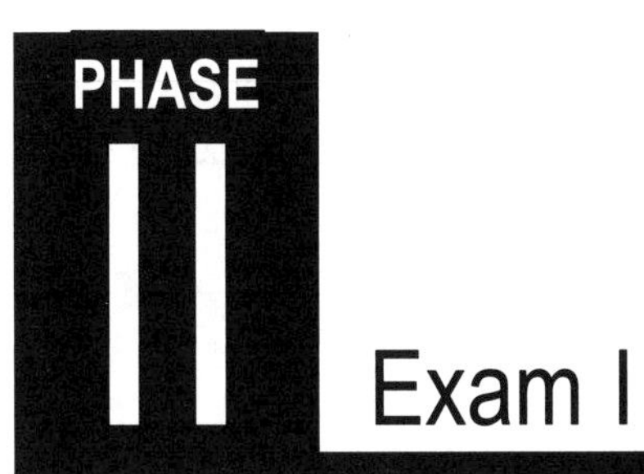

Exam I

1. The National Fire Protection Standard for Fire Officer is ______________.
   a. NFPA 1021
   b. NFPA 1201
   c. NFPA 1001
   d. NFPA 1403
2. A company officer is responsible for the _________ of the assigned personnel in an emergency services organization.
   a. performance
   b. safety
   c. mission development
   d. both a & b
   e. both a & c
3. The company officer spends _____ percent of his/her time dealing with administrative aspects.
   a. 10
   b. 40
   c. 50
   d. 90
4. Which of the following is not a role of the company officer?
   a. coach
   b. counselor
   c. role model
   d. administrator
   e. none of the above
5. Company officers are expected to be _______ to their assigned personnel.
   a. full-time instructor
   b. full-time leader
   c. both a & b
   d. none of the above
6. The principle job of the company officer is to ______________.
   a. write reports
   b. lead others
   c. fight fire
   d. emergency response

7. Professionalism encompasses all of the following except _______________.
   a. attitude
   b. behavior
   c. cognitive
   d. demeanor
   e. ethics

8. A document that attests that one has demonstrated the knowledge and skills necessary to function in a particular craft or trade is known as _______________.
   a. qualification
   b. training
   c. certification
   d. all of the above

9. Communications that usually have legal standing within the organization are known as _______________.
   a. informal
   b. written
   c. informational
   d. formal

10. Communications that are more simple and spontaneous are known as _______________.
    a. formal
    b. informal
    c. written
    d. none of the above

11. The communications model has several steps. The process ends with _______________.
    a. sender
    b. receiver
    c. feedback
    d. medium

12. The area that tends to cause problems in communications is often _______________.
    a. sender
    b. message
    c. receiver
    d. medium

13. An obstacle in communications that prevents the message from being understood by the receiver is a _______________.
    a. barrier
    b. reading
    c. active listening
    d. none of the above

14. Big words, technical terms, and language differences are examples of _______________.

    a. mediums
    b. barriers
    c. understanding
    d. problems

15. The deliberate and apparent process by which one focuses attention on the communications of another is known as _______________.

    a. listening
    b. active listening
    c. passive listening
    d. aggressive listening

16. Which of the following is a part of effective writing skills?

    a. consider the reader
    b. emphasis
    c. brevity
    d. simplicity
    e. all of the above

17. A graphic representation of what the organization should look like and defines the formal lines of authority and responsibility is known as a(n) _______________.

    a. organizational chart
    b. organizational growth
    c. informal organization
    d. group chart

18. A public fire department is part of which of the following?

    a. community
    b. fire district
    c. organizational development
    d. local government

19. A characteristic or organizational structures denoting the relationship between supervisors and subordinates is called _______________.

    a. accountability
    b. line authority
    c. responsibility
    d. flow chart

20. The organizational concept that refers to the uninterrupted series of steps or layers within an organization refers to which of the following?
    a. linear principle
    b. unity principle
    c. scaler principle
    d. division of labor principle

21. The organizational or command principle whereby there is only one boss no matter how many different organizations, divisions, or groups is referred to as _______________.
    a. unity of command
    b. command
    c. area command
    d. multiple command

22. An organizational principle that addresses the number of personnel a supervisor can effectively manage is the _______________.
    a. command
    b. unity of command
    c. delegation
    d. span of control

23. The act of assigning duties to subordinates is _______________.
    a. command
    b. entrepreneurship
    c. delegation
    d. dumping

24. A traditional approach to organizing a company may be characterized where the company officer is reported to by all of his crew. This is known as which type of structure?
    a. linear organizational
    b. straight organizational
    c. flat organizational
    d. round organizational

25. The act of guiding the human and physical resources of an organization to attain an organization's objectives is _______________.
    a. management
    b. command
    c. group dynamics
    d. team building

26. The father of professional management was ______________.
    a. McGregor
    b. Brunacini
    c. Drucker
    d. Fayol
27. Which of the following is the first step in the management process that involves looking to the future and determining objectives?
    a. organizing
    b. planning
    c. commanding
    d. coordinating
28. The final step in the management process involves monitoring the process to ensure that the work is accomplishing the intended goals and objectives and taking corrective action when it is not. This is known as ______________.
    a. coordinating
    b. organizing
    c. controlling
    d. commanding
29. William Ouchi developed the management style in which the manager believes that people not only like to work and can be trusted, but that they want to be collectively involved in the management process and recognized when successful. This is known as which of the following theories?
    a. Theory A
    b. Theory X
    c. Theory Z
    d. Theory Y
30. The management theory on which the organization focuses is the continuous improvement geared to customer satisfaction. This is known as ______________.
    a. total quality management
    b. Theory A
    c. Theory Z
    d. Theory X
31. A formal statement that defines a course or method of action is a ______________.
    a. procedure
    b. policy
    c. plan
    d. program

32. A system of values or a standard of conduct is referred to as _______________.
    a. laws
    b. standards
    c. ethics
    d. principle

33. A formal document indicating the focus, direction, and values of an organization is _______________.
    a. vision
    b. mission statement
    c. SOP
    d. rules and regulations

34. This is a target or other objective by which achievement can be measured. This also helps define purpose and mission.
    a. goal
    b. objective
    c. vision
    d. value

35. A financial plan to purchase high-dollar items that have a life expectancy of more than one year is referred to as a(n) _______________ budget.
    a. operating
    b. line-item
    c. capital
    d. program

36. A financial plan to acquire the goods and services needed to run an organization for a specific period of time is referred to as the _______________ budget.
    a. line-item
    b. program
    c. capital
    d. operating

37. Collecting similar items into a single account and presenting them in a general area collectively under this topic or one line in a budget document is _______________ budget.
    a. capital
    b. operating
    c. program
    d. line-item

38. Which of the following are proven time eaters?
   a. lack of personal goals and objectives
   b. good planning
   c. procrastination
   d. a & c
   e. b & c

39. Which of the following is not one of the three major components of customer service?
   a. Constantly be selective of the services.
   b. Always be nice.
   c. Regard everyone as a customer.
   d. Constantly raise the bar.

40. Activities that take about as long as the time is allowed is known as a concept called the _______________.
   a. Managerial Law
   b. Parkinson's Law
   c. Ranle Law
   d. Empowering Law

41. A group of people working together to accomplish a task is a(n) _______________.
   a. leadership
   b. group
   c. organization
   d. none of the above

42. The personal actions of managers and supervisors to get subordinates to carry out certain actions is known as _______________.
   a. leadership
   b. organization
   c. supervisor
   d. a & c

43. A company officer is considered a _______________.
   a. first-line supervisor
   b. second-line supervisor
   c. both a & b
   d. none of the above

44. A five-tiered representation of human needs that was developed to understand human behavior is known as _______________.
   a. Rorie's Hierarchy of Needs
   b. Maslow's Hierarchy of Needs
   c. Rorie's Anarchy of Needs
   d. Maslow's Anarchy of Needs

45. Hygiene factors keep people satisfied with their work environment was the principle concept of which of the following individuals?
    a. Maslow
    b. Herzberg
    c. Rorie
    d. Pavlow

46. Factors that are regarded as work incentives such as recognition and the opportunity to achieve personal goals is known as _______________.
    a. motivators
    b. demotivators
    c. power
    d. actions

47. A recognition of authority derived from the government or other appointing agency is known as _______________ power.
    a. legitimate
    b. earned
    c. reward
    d. punishment

48. Which of the following is not a leadership style?
    a. directing
    b. consulting
    c. supporting
    d. delegating
    e. advocating

49. To disturb, torment or pester is _______________.
    a. an officer's responsibility
    b. officer's right
    c. harassment
    d. fun

50. The quality of being different is known as _______________.
    a. strange
    b. weird
    c. civil
    d. diversity

51. A person who helps a team member or subordinate improve their knowledge, skills, and abilities is known as a _______________.
    a. counselor
    b. coach
    c. director
    d. chief

52. Which of the following is not one of the four career development roles of a leader?
    a. coach
    b. apprentice
    c. adviser
    d. referral agent

53. An administrative process whereby an employee is punished for not conforming to the organization's rules and regulations is known as _______________.
    a. acceptable
    b. commendation
    c. disciplinary action
    d. grievance

54. The first step in the formal disciplinary process is a(n) _______________.
    a. disciplinary action
    b. written reprimand
    c. oral reprimand
    d. suspension

55. To give authority or power to another is _______________.
    a. empowerment
    b. powerment
    c. leadership
    d. termination

56. A formal dispute between employee and employers over some condition or conditions of employment is a(n) _______________.
    a. ominous sign
    b. conflict
    c. complaint
    d. grievance

57. A disagreement, quarrel, or struggle between two individuals or groups is known as _______________.
    a. complaint
    b. grievance
    c. gripe
    d. conflict

58. A step in the disciplinary process that provides the employee a fresh start in another venue results in a _______________.
    a. transfer
    b. dismissal
    c. commendation
    d. empowerment

59. In 2002, the leading cause of death to firefighters was _______________.
    a. responding/returning from alarms
    b. training
    c. fire ground
    d. non-fire emergencies

60. Firefighter deaths by nature was topped by _____ in 2002?
    a. asphyxiation
    b. heart attacks
    c. burns
    d. internal trauma
    e. b & d

61. Safety is the concern of _______________.
    a. the company officer
    b. everyone
    c. the safety officer
    d. the fire chief
    e. all of the above

62. A person assigned as manager of the department's health and safety program is known as the _______________.
    a. company officer
    b. chief
    c. health and safety director
    d. health and safety officer

63. A team or company of emergency personnel kept immediately available for the potential rescue of other emergency responders is known as the _______________.
    a. safety team
    b. rapid response team
    c. rapid intervention team
    d. rapid safety team

64. An opportunity to take a short break from firefighting duties to rest, cool off, and rehydrate is known as _______________.
    a. rehabilitation
    b. reconciliation
    c. rest period
    d. crew rotation

65. Which NFPA standard states that fire departments should have a physical fitness program?
   a. 1001
   b. 1002
   c. 1021
   d. 1500

66. Which of the following are not generally accepted tools for dealing with critical incident stress?
   a. training
   b. debriefing
   c. alcohol
   d. counseling
   e. scene management

67. An action taken to control and extinguish fire is known as fire _______________.
   a. suppression
   b. prevention
   c. attack
   d. control

68. America Burning was published in _______________.
   a. 1903
   b. 1943
   c. 1973
   d. 1993

69. The actions taken to prevent a fire from occurring or if one occurs to prevent a loss is known as _______________.
   a. fire clean up
   b. fire attack
   c. fire prevention
   d. public education

70. The leading cause of fire-related fatalities in the residential environment is _______________.
   a. arson
   b. smoking
   c. weather
   d. heating appliances

71. Solutions to the nation's fire problem rest with ongoing fire prevention activities. Which of the following is not a part of the three Es of fire prevention?
    a. education
    b. engineering
    c. enforcement
    d. environment

72. Self-operating thermosensitive device that releases a spray of water over a designed area to control or extinguish a fire is known as _______________.
    a. deluge gun
    b. stand pipe
    c. prepiped system
    d. automatic fire protection sprinkler

73. The second stage of fire is known as _______________ phase.
    a. incipient
    b. free-burning
    c. backdraft
    d. smoldering

74. A dramatic event in a room fire that rapidly leads to full involvement of all combustible materials present is _______________.
    a. backdraft
    b. rollover
    c. flashpoint
    d. flashover

75. A mental assessment of the situation, gathering and analyzing information that is crucial to the outcome of the event, is _______________.
    a. pre-planning
    b. pre-incident analysis
    c. size-up
    d. none of the above

76. Which of the following is the leading cause of injuries in fires?
    a. matches
    b. heating equipment
    c. cooking
    d. natural causes

77. Company officers should remember that they have a right to complete an investigation, and that they should have continuous presence on the property until that investigation is complete or an investigative team has released them. Once the premises is vacated, what has to be secured prior to reentering the premises to do an investigation?
    a. probable cause
    b. search warrant
    c. nothing has to occur
    d. administrative process

78. Locations where there are unusual hazards, or where an incident would likely overload the department's resources or where there is a need for interagency cooperation to mitigate the hazard is known as _______________.
    a. hazardous material sites
    b. universities
    c. target hazards
    d. major hazards

79. A bird's-eye view of property showing existing structures for purpose of pre-incident planning is known as a _______________ plan.
    a. floor
    b. plot
    c. life safety
    d. map

80. A portion of a structure that is relatively safe from fire and the products of combustion typically used for protecting occupants in place is known as _______________.
    a. floor area
    b. safe haven
    c. holding area
    d. area of refuge

81. An organizational directive that establishes a standard course of action is called a _______________.
    a. standard operating guideline
    b. standing operating procedure
    c. both a & b
    d. none of the above

82. The NFPA standard for Firefighter Professional Qualifications is _______________.
    a. 1001
    b. 1002
    c. 1021
    d. 1710

83. The NFPA standard for live fire training is _______________.
    a. 1001
    b. 1002
    c. 1017
    d. 1403
84. Significant points in the emergency event usually marking the accomplishments of one of three incident priorities is called _______________.
    a. benchmarking
    b. time marking
    c. trend marking
    d. marc
85. Incident priorities include which of the following?
    a. life safety
    b. incident stabilization
    c. property conservation
    d. all of the above
86. The command sequence consists of which of the following? 1. life safety 2. developing an action plan 3. implementing the action plan 4. pre-incident planning 5. post-incident analysis
    a. 1, 3, 4
    b. 1, 2, 4
    c. 1, 2, 3
    d. all of the above
87. Firefighting operations that make a direct attack on the fire for purposes of control and extinguishment is _______________ mode.
    a. defensive
    b. transitional
    c. command
    d. offensive
88. This sets broad goals and outlines the overall plan to control the incident.
    a. tactics
    b. strategy
    c. size-up
    d. action plan
89. The various maneuvers that can be used to achieve a strategy while fighting a fire or dealing with a similar emergency is known as _______________.
    a. tactics
    b. strategy
    c. size-up
    d. action plan

90. The strategically combined various maneuvers that are used to achieve a strategy while fighting a fire or dealing with a similar emergency is known as _______________.
    a. tactics
    b. strategy
    c. size-up
    d. action plan

91. The critical process of shifting from the offensive mode to the defensive mode or the defensive mode to the offensive mode is known as _______________ mode.
    a. switching
    b. moving
    c. transitional
    d. flip flop

92. The on-going process of evaluating the incident throughout its duration is known as _______________.
    a. strategizing
    b. tactic deployment
    c. action planning
    d. size-up

93. Which of the following is a legal document that sets forth requirements to protect health, safety, and the general welfare of the public as they relate to construction and occupancy of buildings and construction?
    a. fire prevention codes
    b. building codes
    c. city codes
    d. fire codes

94. The typical downtown business districts in older communities where the buildings are generally no more than five stories tall and the exterior is masonry is generally known as which type of building construction?
    a. Type I
    b. Type II
    c. Type III
    d. Type IV

95. As a fire continues to build the smoke levels drop. When hot, fuel-rich gases meet fresh air, they ignite, and burn along the ceiling. This is known as _______________.
    a. backdraft
    b. flashover
    c. combustion
    d. rollover

96. Which of the following is not a building construction type?
    a. ordinary
    b. heavy roof
    c. non-combustible
    d. heavy timber

97. An assessment of the safety hazards for both civilians and firefighters in a particular occupancy with lightweight truss construction is known as the _______________ factors.
    a. survival
    b. enhancement
    c. structural
    d. occupancy

98. Fire officers must base their actions off of priorities such as life safety, property conservation, and incident stabilization. Which of the following puts these in the correct order?
    a. life safety, incident stabilization, property conservation
    b. incident stabilization, life safety, property conservation
    c. property conservation, life safety, incident stabilization
    d. property conservation, incident stabilization, life safety

99. The act of controlling the fire to prevent the fire from extending any further in the structure is known as _______________.
    a. fire detection
    b. overhaul
    c. extinguishment
    d. confinement

100. The difference between the fire triangle and the fire tetrahedron is _______________.
    a. heat
    b. fuel
    c. chemical chain reaction
    d. oxygen

Phase II, Exam I: Answers to Questions

1. A
2. D
3. D
4. D
5. A
6. B
7. C
8. C
9. D
10. B
11. C
12. D
13. A
14. B
15. B
16. E
17. A
18. D
19. B
20. C
21. A
22. D
23. C
24. C
25. A
26. D
27. B
28. C
29. C
30. A
31. B
32. C
33. B
34. A
35. C
36. D
37. D
38. D
39. A
40. B
41. C
42. A
43. A
44. B
45. B
46. A
47. A
48. E
49. C
50. D
51. B
52. B
53. C
54. C
55. A
56. D
57. D
58. A
59. C
60. B
61. E
62. D
63. C
64. A
65. D
66. C
67. A
68. C
69. C
70. A
71. D
72. D
73. B
74. D
75. C
76. C
77. B
78. C
79. B
80. D
81. B
82. A
83. D
84. A
85. D
86. C
87. D
88. B
89. A
90. D
91. C
92. D
93. B
94. C
95. D
96. B
97. A
98. A
99. D
100. C

Phase II, Exam I: Rationale & References for Questions

Question #1. It is important to know the standard from which the program and requirements are coming from. NFPA 1021 1.1. *CO, 2E,* Page 9.

Question #2. The role of the company officer is to manage personnel at the crew level. Performance and safety are key items for which the officer is responsible. NFPA 4.2.1. *CO, 2E,* Page 6 & 7.

Question #3. The company officer is responsible for many items. Operations account for 10 percent and administrative aspects 90 percent, including doing reports, evaluations, and other administrative document record-keeping. NFPA 4.2.1, 4.2.2. *CO, 2E,* Page 5.

Question #4. The roles of company officers are many. It is important to know your roles as an officer. Effective company officers fill many roles throughout their careers. Many are done simultaneously. NFPA 4.2. *CO, 2E,* Page 5.

Question #5. These two pieces are some of the most important parts of what company officers do in a daily function. Human relations are extremely important to the crews and the success of the company officer. NFPA 4.2.2 ; 4.2.3; 5.2.1; 5.2.2. *CO, 2E,* Page 5.

Question #6. Leading others is the company officer's principle job. The capabilities, efficiency, and morale of the company are direct reflections of the company officer's leadership abilities. NFPA 5.2; 4.2. *CO, 2E,* Page 5.

Question #7. Being a role model means being professional. It is important to have all of the necessary qualities to be an effective leader. NFPA 4.2; 5.2. *CO, 2E,* Page 15.

Question #8. Certification means that an individual has been tested by an accredited examining body on clearly identified material and found to meet the minimum standard. NFPA 3.3.11. *CO, 2E,* Page 8.

Question #9. Good communication skills are essential in work and personal life. Formal communications are conducted according to established standards. They tend to follow customs, rules, and practices. NFPA 4.2.2; 5.2.1. *CO, 2E,* Page 20.

Question #10. We write with less formality when sending emails, memos, and short notes. We also tend to be less formal at social events. NFPA 5.2.1; 4.2.1; 4.2.2. *CO, 2E,* Page 20.

Question #11. This is the final step in the communications process. NFPA 4.3.3; 4.2.2; 5.2. *CO, 2E,* Page 21.

Question #12. Mediums can be a variety of problems to include noise, language, terminology, and others. It is important to try to control this area as much as possible so your communications will be as effective and as clear as possible. NFPA 4.3.3; 4.2.1; 5.2.1; 5.2.2. *CO, 2E,* Pages 21.

Question #13. Consider a barrier to be like a filter. One or more filters reduces the information flow between the sender and the receiver. NFPA 4.2; 5.2. *CO, 2E,* Pages 23.

Question #14. There are many barriers to effective communications. These are some of the common ones. NFPA 5.2; 4.2. *CO, 2E,* Page 24.

Question #15. In active listening, the listener can actually show the sender that you are actively listening by focusing all of your attention on the speaker and showing a genuine interest in the message. Active listening is a significant portion of good communications. NFPA 5.2; 4.2. *CO, 2E,* Page 24.

Question #16. The importance of written communications is imperative. Many of these documents will be official records and will also serve as information being sent to the public. NFPA 4.3.3; 5.2.2; 5.2.1. *CO, 2E,* Page 28.

Question #17. An organizational chart defines the roles and lines of authority that are important in a paramilitary organizational structure. It is important to know where you fall in the organizational chart as a company officer. NFPA 4.4; 4.6; 5.4; 5.6. *CO, 2E,* Page 47 & 48.

Question #18. It is important to understand the infrastructure of the community and government of which the fire department is a part. NFPA 4.3.1. *CO, 2E,* Page 54.

Question #19. An officer with line authority manages one or more of the functions that are essential for the fire department's mission. When we see an organizational chart, we usually think of the authority one has and in the areas they have it. NFPA 4.4.2; 5.2.1. *CO, 2E,* Page 62.

Question #20. Like playing a scale on a musical instrument where every note is sounded, the scaler principle suggests that every level in the organization is considered in the flow of communications. NFPA 4.4.2. *CO, 2E,* Page 66.

Question #21. Reporting to one boss is the concept in this portion. Unity of command is an essential organizational concept. NFPA 4.2.2; 5.2.1. *CO, 2E,* Page 66.

Question #22. The number of people you can effectively supervise varies of course, based on many factors, but generally for our purposes the number is between four and seven. NFPA 5.2.1; 5.6.1. *CO, 2E,* Page 67.

Question #23. As an organization expands, we assign certain duties to subordinates. It is important that delegation be mutually understood and that authority be provided along with the responsibility. NFPA 4.2.2. *CO, 2E,* Page 67.

Question #24. This organizational structure follows some of the rules about flat and lean organizations and allows for good communications between the company officer and all of the firefighters on the crew. NFPA 4.2.2. *CO, 2E,* Page 69.

Question #25. The act of guiding the human and physical resources of an organization to attain the organization's objectives. The company officer is the key in this principle where as they are the ones managing the work force. NFPA 4.2.1; 5.2.1. *CO, 2E,* Page 75.

Question #26. Fayol is known as the father of management. He is known for his research and development of managerial principles. NFPA 5.2.1. *CO, 2E,* Page 77.

Question #27. Planning covers everything from the next hour to the next decade. NFPA 4.2.3. *CO, 2E,* Page 77.

Question #28. Controlling helps us get to the right place at the right time through the monitoring of efforts of the resources. This is a large part of the company officer's job. NFPA 4.2.1. *CO, 2E,* Page 79.

Question #29. As a company officer, it is important to understand which style of management practices you must choose. Each situation will require a different style or use of theory. NFPA 5.2.1. *CO, 2E,* Page 81.

Question #30. Total quality management is based on employee participation and the concept that all must work together to achieve goals. This is true of the fire service especially on emergency scenes. NFPA 5.2.1. *CO, 2E,* Page 82.

Question #31. Management can be enhanced through good policy, procedures and even by personal observation. Policies are broad in nature and should be clearly understood. NFPA 4.3.1; 5.2.1. *CO, 2E,* Page 84.

Question #32. Ethics often begin where laws leave off. Ethics have a direct impact on the management of the fire service. These are rules set by the profession. NFPA 4.1.1. *CO, 2E,* Page 89.

Question #33. The mission statement declares the vision of the department by setting specific values and setting the focus in which direction the department is moving. A mission statement is like putting up a sign for employees on which direction to go. NFPA 5.2; 4.2. *CO, 2E,* Page 97.

Question #34. Goals help you identify where you are going. These take several years often to accomplish. In a sense, this is a strategic road map with the planned stops along the way to reaching the mission. NFPA 4.4. *CO, 2E,* Page 98.

Question #35. Classic examples of capital budget items are fire stations, apparatus purchase, and any large ticket item. It is important to understand the budgeting process whereas you are managing the money at the first level in protecting the organization's investments. NFPA 4.4.3; 5.4.2. *CO, 2E,* Page 103.

Question #36. This is everything not covered typically in the capital budget. The operating budget is a general budget that shows specific amounts needed to operate the organization in general form. NFPA 5.4.2; 4.4.3. *CO, 2E,* Page 103.

Question #37. Line-item budgeting is like keeping accounting on the money spent. This is used when you need specifics to be purchased and don't want it to be generic in nature. NFPA 4.4.3; 5.4.2. *CO, 2E,* Page 103.

Question #38. People allow important events to become urgent events through the lack of good time management techniques. Putting off work creates rushes and inefficiency. Thus, this produces lower quality work performance that as an end result, hurts the organization. NFPA 4.4.2; 5.2.1. *CO, 2E,* Page 106.

Question #39. Our customers are the tax payers and the employees that work for us. We must address the interior and exterior components of this concept. The three pieces are great guidelines to go by. NFPA 4.3.1; 4.3.3. *CO, 2E,* Page 110 & 111.

Question #40. Managing time is an important aspect of a company officer. Time is the one thing that is hard to track for efficiency based upon the tasks performed and the personnel performing them. It is important, however, not to fall into this concept of allowing the prescribed time to dictate the time needed for the event. NFPA5.2.1; 5.2.2; 5.2.3; 4.4.2. *CO, 2E,* Page 1058.

Question #41. There are many types of organizations. The important part is that it is a group of people working to accomplish a specific set of goals or functions. NFPA 4.2. *CO, 2E,* Page 119.

Question #42. Leadership is accomplishing the organization's goals through others. NFPA 4.4.1. *CO, 2E,* Page 120.

Question #43. The company officers are first-line supervisors because they supervise firefighters. Because they do not supervise other managers, they are considered to be the first-line level of supervision. NFPA 4.6. *CO, 2E,* Page 121.

Question #44. Using models of human behavior helps us understand people and their actions and motivation. NFPA 5.2.1. *CO, 2E,* Page 123.

Question #45. Using models of human behavior helps us understand people and their actions and motivation. NFPA 5.2.1. *CO, 2E,* Page 124 & 125.

Question #46. Motivators encourage employees to rise above the satisfactory level and do excellent work. NFPA 4.2; 5.2.1. *CO, 2E,* Page 125.

Question #47. This is the type of power that comes with the position or the "badge." NFPA 4.2.1. *CO, 2E,* Page 127.

Question #48. Dynamic and effective leaders make their style fit the situation. There is importance in utilizing the different levels based on the individuals and the tasks. NFPA4.2.3; 5.2.2. *CO, 2E,* Page 128.

Question #49. Harassment of any kind should not be allowed in the work place. Title VII of the Civil Rights Act of 1964 NFPA 4.7.1. *CO, 2E,* Page 133.

Question #50. To have our firefighter population meet the needs of the citizens, it is important to have the total community in view. This means to have representation of various cultures working to help create better understanding and service needs to the entire community. NFPA 4.7.1. *CO, 2E,* Page 132.

Question #51. Coaching is an informational process that helps subordinates improve their skills and abilities. Coaching implies one on one relationship that treats subordinates as full partners. NFPA 5.2.1. *CO, 2E,* Page 147.

Question #52. As a leader, you will be in one of these roles with different subordinates at the same time. It is important to understand as the individual grows, your role will change. NFPA 5.2.1. *CO, 2E,* Page 147 -150.

Question #53. Disciplinary action is another way of improving performance and skills. This lets the subordinate know that their performance is below acceptable standards. NFPA 5.2.2. *CO, 2E,* Page 160.

Question #54. This is much like a private counseling session to set direction for the subordinate. NFPA 5.2.2. *CO, 2E,* Page 181.

Question #55. Empowerment allows members to have a feeling of ownership in the organization. This is accomplished by getting them involved. NFPA 5.2.1. *CO, 2E,* Page 162.

Question #56. This is a serious issue and usually occurs when the condition reaches the point that complaining doesn't seem to help fix or solve the issue or bring any relief. NFPA 5.2.2. *CO, 2E,* Page 163.

Question #57. When individuals work together, there will always be friction of dislike by someone for a variety of reasons. This behavior is a conflict. NFPA 5.2.1; 5.2.2. *CO, 2E,* Page 163.

Question #58. This gives the individual a new start and often will allow for changes without embarrassment. NFPA 5.2.2. *CO, 2E,* Page 161.

Question #59. It is important to understand the causes of injuries and death of firefighters to take a proactive approach to reduce these numbers. NFPA 4.7.1; 4.7.2. *CO, 2E,* Page 179.

Question #60. It is important to understand the causes of injuries and death of firefighters to take a proactive approach to reduce these numbers. NFPA 4.7.1; 4.7.2. *CO, 2E,* Page 179.

Question #61. Safety comes from the elimination of unsafe equipment and procedures, having and enforcing policies that provide for safety, and having viable recognition programs at all levels that reinforce both the company's and industry's commitment to worker safety and health. NFPA 4.7.1; 4.7.2; 5.7.1. *CO, 2E,* Page 178.

Question #62. NFPA 1500 states every department should have an individual assigned to these duties. NFPA 4.7.1; 5.7.1. *CO, 2E,* Page 184.

Question #63. This concept is indicated in NFPA 1500 where an emphasis on accountability and safety during emergency operations has significant emphasis. NFPA 4.7. *CO, 2E,* Page 197.

Question #64. Rehabilitation is a significant requirement to assure that the personnel working on an emergency scene are physically able to continue working. This is a way to monitor health conditions and potentially recognize signs that could indicate when firefighters are not medically well to continue working. This concept could potentially lower the firefighter death rate in the heart attack area significantly. NFPA 4.7.1. *CO, 2E,* Page 197.

Question #65. Physical fitness lowers injury and death rates by conditioning firefighters for the job. It is important as an officer to know what standards are in place for safety. NFPA 4.7.1; 4.7.2. *CO, 2E,* Page 202.

Question #66. Critical incident stress is a significant part of an emergency responder's life. Assistance is often necessary. NFPA 4.6.4. *CO, 2E,* Page 206.

Question #67. Fire suppression is a reaction—resources are mobilized following an event and the action to mitigate an event. NFPA 4.6.3. *CO, 2E,* Page 214.

Question #68. NFPA 4.7. *CO, 2E,* Page 214.

Question #69. This is a proactive approach; it involves the activities that help keep fires from occurring. NFPA 5.5. *CO, 2E,* Page 214.

Question #70. Careless use and discarding of smoking materials is the leading cause of fire-related fatalities in the residential setting. NFPA 5.5. *CO, 2E,* Page 222.

Question #71. Life safety education is the first priority in emergency operations. Life safety addresses the safety of occupants and emergency responders. NFPA 4.3.4. *CO, 2E,* Page 222 - 224.

Question #72. NFPA 4.6.1, 5.6.1. *CO, 2E,* Page 254.

Question #73. NFPA 4.6.2, 5.6.2. *CO, 2E,* Page 257.

Question #74. NFPA 4.7.1, 4.6.2, 5.6.2. *CO, 2E,* Page 258.

Question #75. NFPA 4.6.2, 5.6.1. *CO, 2E,* Page 323 - 326.

Question #76. Unsafe practices, poor housekeeping, inappropriate attire, and lack of maintenance can cause fires associated with cooking. NFPA 4.5.1, 5.5.2. *CO, 2E,* Page 273.

Question #77. Fire investigators must work carefully to locate, gather, and preserve evidence. NFPA 4.5.1, 4.5.2 5.5.2. *CO, 2E,* Page 281.

Question #78. Given that you cannot plan for every situation, you plan for those that present the most significant risk. NFPA 4.6.2, 5.6.1. *CO, 2E,* Page 293.

Question #79. Having an overview of exposures and the other hazards are easily identified in this style of preplan. NFPA 4.6.1. *CO, 2E,* Page 295.

Question #80. With rescue being the number one priority in emergency responses, it is important to know pre-designated locations where individuals will go to reach safety during an emergency. NFPA 4.6.3 5.6.1. *CO, 2E,* Page 298.

Question #81. Unlike a pre-plan, an SOP is used to prepare with other situations that are likely to occur in the jurisdiction. Officers are frequently required to recall from memory these standard operating procedures. NFPA 4.7 5.7. *CO, 2E,* Page 304.

Question #82. This standard provides the basic requirements for firefighters in their proficiency at the job requirements. NFPA 4.7.1. *CO, 2E,* Page 308.

Question #83. When acquired structures are used for training, they must be carefully inspected and prepared for the training evolution. This standard provides those guidelines. NFPA 4.7.1. *CO, 2E,* Page 310.

Question #84. These key pieces of communication are utilized by the incident commander to note accomplishment of the three critical priorities of any incident. NFPA 4.6, 4.7, 5.6, 5.7. *CO, 2E,* Page 320.

Question #85. With these priorities, they provide a systematic approach to emergency incidents. NFPA 4.6, 5.6. *CO, 2E,* Page 320.

Question #86. The command sequence is a three-step process that helps the incident commander effectively manage the scene. NFPA 4.6 , 5.6. *CO, 2E,* Page 322.

Question #87. The strategies and tactics selected to control and mitigate an incident are crucial. Selecting the correct mode is important to the safety of all personnel working the incident. NFPA 4.6, 5.6. *CO, 2E,* Page 327.

Question #88. When the strategy is well-defined, all personnel understand the tasks at hand and can focus their efforts on making it happen. NFPA 4.6, 5.6. *CO, 2E,* Page 333.

Question #89. Tactics help you reach your strategic goals. NFPA 4.6 5.6. *CO, 2E,* Page 333.

Question #90. Action plan puts the planning and thinking phases into motion. NFPA 4.6 5.6. *CO, 2E,* Page 334.

Question #91. This is a phase that you will go through as you transition from one phase to another. You would not start out in this mode. NFPA 4.6, 5.6. *CO, 2E,* Page 329.

Question #92. Size-up includes more than what you see through the windshield as you pull up. It is an analytical process that occurs throughout the incident to provide feedback to the incident commander and officers to verify if their strategies and tactics are working. NFPA 4.6, 5.6. *CO, 2E,* Page 323.

Question #93. An important part of fire prevention and the company officer's job is to enforce codes that are applicable to the jurisdiction. Many of these cross between building and fire prevention. It is important to understand how each relates. NFPA 5.5.1. *CO, 2E,* Page 228.

Question #94. Action plans for fire attack are based upon fire behavior and the building construction to determine the appropriate tactics. Knowing the correct building construction will help establish strategic goals to mitigate the incident. NFPA 4.6.1. *CO, 2E,* Page 249.

Question #95. It is important to understand the growth and progression of fire to aid in developing action plans. NFPA 4.6.2. *CO, 2E,* Page257 - 258.

Question #96. Building construction will define fire behavior and the tactics required to extinguish a fire. NFPA 4.6.2. *CO, 2E,* Page 248 - 251.

Question #97. Stairwells and other penetrations to allow for rescue, fire spread, and potential falling hazards for firefighters. These are important to company officers as they work and direct crews in tactical operations. NFPA 4.6.2. *CO, 2E,* Page 248.

Question #98. The incident commander and the company officer are both responsible for these items whereas the initial arriving company officer is the incident commander for at least a few minutes. These priorities are how action plans are formulated. NFPA 5.6.1. *CO, 2E,* Page 321.

Question #99. Benchmarks or progression in fire attack is important to the company officer as definitive positions in the incident. NFPA 4.6.3. *CO, 2E,* Page 335.

Question #100. Understanding fire growth is important as a company officer as you make multiple split-second decisions on tactics and operations based upon the knowledge of fire behavior. NFPA 4.6.2. *CO, 2E,* Page 257.

Exam II

1. Persons with high affiliation needs don't like to work in groups.
   a. True
   b. False
2. Multi-company training evolutions provide an opportunity to exercise the Incident Management System.
   a. True
   b. False
3. Regardless of whether the firefighter is a career employee or a volunteer member, injuries cost money.
   a. True
   b. False
4. In most departments, companies spend less than 10 percent of their time dealing with non-emergencies.
   a. True
   b. False
5. Actions taken during the first five minutes of an incident have a significant impact on the overall outcome of the incident.
   a. True
   b. False
6. Good time management correlates the management of our time with previously set goals and objectives.
   a. True
   b. False
7. Strong leadership and closed communications are vital to the success of any organization.
   a. True
   b. False
8. Pre-incident planning is the fact-finding part of the pre-incident planning process.
   a. True
   b. False
9. The cycle of performance management represents the continuous process of goal setting, observation, and performance evaluation.
   a. True
   b. False

10. Career development is a shared responsibility. Both the firefighter and the department have an obligation to this growth.
    a. True
    b. False
11. The crime of arson requires criminal intent as well as the act itself.
    a. True
    b. False
12. Experienced fire officers should be able to determine the origin and cause of nearly all the fires they attend.
    a. True
    b. False
13. Planning does not have to start with a clear understanding of the goals and objectives.
    a. True
    b. False
14. Persons with low power needs like to be in charge.
    a. True
    b. False
15. Nearly half of firefighters killed during structural firefighters' operations were advancing hose lines at the time of their injury.
    a. True
    b. False
16. Persons with low achievement needs like challenges.
    a. True
    b. False
17. Your pre-incident planning should not address the community's overall risk.
    a. True
    b. False
18. A fire department's Fire Suppression Rating Schedule is directly related to the rating classification given by the Insurance Service Office.
    a. True
    b. False
19. Lightweight truss construction is primarily used in Type I construction.
    a. True
    b. False
20. Effective fire officers do not have the need to write effectively.
    a. True
    b. False

21. These factors keep people satisfied with their work environment.
    a. motivating factors
    b. hygiene factors
    c. responsibility factors
    d. none of the above
22. This duty involves activities that will identify hazards in occupancies.
    a. pre-plans
    b. pre-fire surveys
    c. inspections
    d. pre-incident analysis
23. A legal term denoting deliberate and unlawful burning of property is known as _______________.
    a. fire
    b. arson
    c. incendiary devices
    d. abuse
24. A characteristic of organizational structures denoting the relationship between supervisors and subordinates is known as _______________.
    a. line of authority
    b. accountability
    c. organization theory
    d. group dynamics
25. A characteristic or organizational structures denoting the relationship between supervisors and subordinates is called _______________.
    a. accountability
    b. line authority
    c. responsibility
    d. flow chart
26. The act to disturb, torment, or pester means _______________.
    a. bug
    b. aggravate
    c. harassment
    d. none of the above
27. The act of being accountable for actions and activities along with having a moral and perhaps legal obligation to carry out certain activities is known as _______________.
    a. accountability
    b. responsibility
    c. line of authority
    d. all of the above

28. A formal document indicating the focus and values for an organization is known as the _______________.
    a. vision
    b. introduction
    c. mission statement
    d. rules and regulations

29. A recognition of authority by virtue of the other individual's character or trust is _______________power.
    a. expert
    b. direct
    c. punishment
    d. identification

30. Activities that take about as long as the time is allowed is known as a concept called the _______________.
    a. Managerial Law
    b. Parkinson's Law
    c. Ranle Law
    d. Empowering Law

31. The company officer spends _____ percent of their time dealing with administrative aspects.
    a. 10
    b. 40
    c. 50
    d. 90

32. A type of building construction in which the exterior walls are usually made of masonry and the interior of non-combustibles is known as _______________.
    a. Type I
    b. Type II
    c. Type III
    d. Type IV

33. The organizational or command principle whereby there is only one boss no matter how many different organizations, divisions, or groups is referred to as _______________.
    a. unity of command
    b. command
    c. area command
    d. multiple command

34. Which of the following are not generally accepted tools for dealing with critical incident stress?
    a. training
    b. debriefing
    c. alcohol
    d. counseling
    e. scene management

35. The first step in the formal disciplinary process is a(n) ______________.
    a. disciplinary action
    b. written reprimand
    c. oral reprimand
    d. suspension

36. A recognition of authority derived from the government or other appointing agency is _______________ power.
    a. demanded
    b. granted
    c. legitimate
    d. reward

37. Which of the following is the deliberate and apparent process by which one focuses attention on the communications of another?
    a. oral communications
    b. written communications
    c. active listening
    d. passive listening

38. What is the leading cause of civilian fire deaths?
    a. cooking
    b. heating appliances
    c. children
    d. careless smoking

39. Which of the following is a legal document that sets forth requirements to protect health, safety, and the general welfare of the public as they relate to construction and occupancy of buildings and construction?
    a. fire prevention codes
    b. building codes
    c. city codes
    d. fire codes

40. The act of guiding the human and physical resources of an organization to attain an organization's objectives is _______________.
    a. management
    b. command
    c. group dynamics
    d. team building
41. A formal statement that defines a course or method of action is known as _______________.
    a. procedure
    b. protocol
    c. policy
    d. none of the above
42. The second stage of fire is known as _______________ phase.
    a. incipient
    b. free-burning
    c. backdraft
    d. smoldering
43. Company officers are referred to as _______________.
    a. first-line supervisors
    b. up-line supervisors
    c. equals
    d. none of the above
44. A person assigned as the manager of the department's health and safety program is the _______________ officer.
    a. safety
    b. health
    c. health and safety
    d. division of life safety
45. An opportunity to take a short break from firefighting duties to rest, cool off, and rehydrate is known as _______________.
    a. rehabilitation
    b. reconciliation
    c. rest period
    d. crew rotation
46. In the elements of communications process, the model starts with which of the following?
    a. feedback
    b. sender
    c. receiver
    d. medium

47. The strategically combined various maneuvers that are used to achieve a strategy while fighting a fire or dealing with a similar emergency is known as _______________.
    a. tactics
    b. strategy
    c. size-up
    d. action plan

48. The method of transmission in the communications model is known as the_________.
    a. receiver
    b. message
    c. medium
    d. feedback

49. The NFPA standard for live fire training is _______________.
    a. 1001
    b. 1002
    c. 1017
    d. 1403

50. In the communications model, the information being sent to another is known as _______________.
    a. message
    b. feedback
    c. medium
    d. none of the above

51. America Burning was published in _______________.
    a. 1903
    b. 1943
    c. 1973
    d. 1993

52. This sets broad goals and outlines the overall plan to control the incident.
    a. tactics
    b. strategy
    c. size-up
    d. action plan

53. A labor relations term denoting that union membership is a condition of employment is _______________ shop.
    a. union
    b. closed
    c. open
    d. agency

54. A financial plan to purchase high dollar items that have a life expectancy of more than one year is a(n) _______________ budget.
    a. line-item
    b. program
    c. capital
    d. operating

55. Self-operating thermosensitive device that releases a spray of water over a designed area to control or extinguish a fire is known as a(n) _______________.
    a. deluge gun
    b. stand pipe
    c. prepiped system
    d. automatic fire protection sprinkler

56. The NFPA standard for firefighter professional qualifications is _______________.
    a. 1001
    b. 1002
    c. 1021
    d. 1710

57. The quality of being different is known as _______________.
    a. strange
    b. weird
    c. civil
    d. diversity

58. Company officers should remember that they have a right to complete an investigation, and that they should have continuous presence on the property until that investigation is complete or an investigative team has released them. Once the premises is vacated, what has to be secured prior to reentering the premises to do an investigation?
    a. probable cause
    b. search warrant
    c. nothing has to occur
    d. administrative process

59. This is a graphic representation of what the organization should look like. It defines the formal lines of authority and responsibility.
    a. organizational chart
    b. organizational growth
    c. informal organization
    d. group chart

60. Which NFPA standard states that fire departments should have a physical fitness program?

    a. 1001

    b. 1002

    c. 1021

    d. 1500

61. The traditional organizational structure design where each individual member of the crew answers directly to the officer is known as _______________ organizational structure.

    a. flat

    b. round

    c. scaler

    d. pyramid

62. Diversity is a quality of being _______________.

    a. diverse

    b. different

    c. all alike

    d. a & b

    e. b & c

63. A financial plan to acquire the goods and services to run an organization for a specific period of time is a(n) _______________ budget.

    a. capital

    b. operating

    c. line-item

    d. program

64. A group of people working together to accomplish a task is known as a _______________.

    a. pool

    b. task force

    c. group

    d. organization

65. The easiest and most common form of communication is _______________.

    a. written

    b. oral

    c. radio

    d. email

66. The personal actions of managers and supervisors to get subordinates to carry out certain actions is known as _______________.
    a. leadership
    b. organization
    c. supervisor
    d. a & c

67. The purpose of the disciplinary process is to improve the subordinate's _______________.
    a. performance
    b. conduct
    c. both a & b
    d. none of the above

68. A person who helps a team member or subordinate improve their knowledge, skills, and abilities is known as a _______________.
    a. counselor
    b. coach
    c. director
    d. chief

69. A mental assessment of the situation; gathering and analyzing information that is crucial to the outcome of the event is _______________.
    a. pre-planning
    b. pre-incident analysis
    c. size-up
    d. none of the above

70. To disturb, torment, or pester is _______________.
    a. an officer's responsibility
    b. an officer's right
    c. harassment
    d. fun

71. The critical process of shifting from the offensive mode to the defensive mode or the defensive mode to the offensive mode is known as _______________ mode.
    a. switching
    b. moving
    c. transitional
    d. flip flop

72. As a fire continues to build, the smoke levels drop. When hot, fuel-rich gases meet fresh air, they ignite, and burn along the ceiling. This is known as _______________.

    a. backdraft
    b. flashover
    c. combustion
    d. rollover

73. The deliberate and apparent process by which one focuses attention on the communications of another is known as _______________.

    a. listening
    b. active listening
    c. passive listening
    d. aggressive listening

74. The five tiered representation of human needs developed by Abraham Maslow is known as Maslow's _______________.

    a. Sequence of Needs
    b. Directional Needs
    c. Hierarchy of Needs
    d. Hierarchy of Desires

75. The activity of providing emergency services is known as a _______________ function.

    a. staff
    b. non-essential
    c. support
    d. line

76. The final step in the management process involves monitoring the process to ensure that the work is accomplishing the intended goals and objectives and taking corrective action when it is not. This is known as _______________.

    a. coordinating
    b. organizing
    c. controlling
    d. commanding

77. Total quality management or TQM is was developed by _______________.

    a. Maslow
    b. Ouchi
    c. Drucker
    d. Deming

78. The providing of human resources in paid or volunteer departments is known as _______________.
   a. fire department
   b. public safety
   c. staffing
   d. jurisdiction

79. An effective tool in employee development that is described as helping a team member improve his knowledge, skills, and abilities is known as _______________.
   a. counseling
   b. evaluation
   c. coaching
   d. all of the above

80. The NFPA Standard for Fire Officer is _______________.
   a. 1001
   b. 1002
   c. 1021
   d. 1020

81. A formal statement that defines a course or method of action is a _______________.
   a. procedure
   b. policy
   c. plan
   d. program

82. A document that attests that one has demonstrated the knowledge and skills necessary to function in a particular craft or trade is known as _______________.
   a. qualification
   b. training
   c. certification
   d. all of the above

83. A group of people working together to accomplish a task is a(n) _______________.
   a. leadership
   b. group
   c. organization
   d. none of the above

84. The concept of an organizational structure utilized on large emergency scenes whereby there is only one boss is known as _______________.
   a. command
   b. operations
   c. unity of command
   d. division of labor

85. A portion of a structure that is relatively safe from fire and the products of combustion typically used for protecting occupants in place is known as _______________.
    a. floor area
    b. safe haven
    c. holding area
    d. area of refuge

86. The report of a study published in 1973 by the National Commission on Fire Prevention and control is known as _______________.
    a. America Did It
    b. America Burning
    c. America Burning Revisited
    d. Councils Report on the Nation's Fire Problem

87. The organizational concept that refers to the uninterrupted series of steps or layers within an organization is known as _______________.
    a. scaler organization
    b. unity of command
    c. linear organization
    d. flat organization

88. A systematic arrangement of a body of rules is _______________.
    a. standard
    b. NFPA
    c. skills
    d. codes

89. The fourth step in the management process which involves the manager's directing and overseeing the efforts of others is _______________.
    a. coordinating
    b. commanding
    c. organizing
    d. controlling

90. A company officer is what level of supervisor?
    a. up-line
    b. down-line
    c. first
    d. second

91. A bird's-eye view of property showing existing structures for purposes of pre-incident planning is known as a _______________ plan.
    a. floor
    b. plot
    c. life safety
    d. map

92. The act of assigning duties to subordinates is known as ______________.
    a. advocation
    b. delegation
    c. administration
    d. relocation

93. Which of the following is an obstacle in the communications process?
    a. physical barrier
    b. personal barrier
    c. semantic barrier
    d. all of the above

94. The first step in the management process which involves looking into the future and determining objectives is known as ______________.
    a. organizing
    b. planning
    c. strategies
    d. tactics

95. In 2002, the leading cause of death to firefighters was ______________.
    a. responding/returning from alarms
    b. training
    c. fire ground
    d. non-fire emergencies

96. The command sequence consists of which of the following? 1. life safety 2. developing an action plan 3. implementing the action plan 4. pre-incident planning 5. post-incident analysis
    a. 1, 3, 4
    b. 1, 2, 4
    c. 1, 2, 3
    d. all of the above

97. Law or regulation that establishes minimum requirements for the design and construction of a building is ______________.
    a. fire code
    b. inspections code
    c. building code
    d. permits

98. Professionalism encompasses all of the following except ______________.
    a. attitude
    b. behavior
    c. cognitive
    d. demeanor
    e. ethics

99. The actions taken to prevent a fire from occurring or if one occurs to prevent a loss is known as _______________.

    a. fire clean up
    b. fire attack
    c. fire prevention
    d. public education

100. An assessment of the safety hazards for both civilians and firefighters in a particular occupancy with lightweight truss construction is known as the _______________ factors.

    a. survival
    b. enhancement
    c. structural
    d. occupancy

Phase II, Exam II: Answers to Questions

1. F
2. T
3. T
4. F
5. T
6. T
7. F
8. T
9. T
10. T
11. T
12. T
13. F
14. F
15. T
16. F
17. F
18. T
19. F
20. F
21. B
22. C
23. B
24. A
25. B
26. C
27. B
28. C
29. D
30. B
31. D
32. B
33. A
34. C
35. C
36. C
37. C
38. D
39. B
40. A
41. C
42. B
43. A
44. C
45. A
46. B
47. D
48. C
49. D
50. A
51. C
52. B
53. B
54. C
55. D
56. A
57. D
58. B
59. A
60. D
61. A
62. D
63. B
64. D
65. B
66. A
67. A
68. B
69. C
70. C
71. C
72. D
73. B
74. C
75. D
76. C
77. D
78. C
79. C
80. C
81. B
82. C
83. C
84. C
85. D
86. B
87. A
88. D
89. A
90. C
91. B
92. B
93. D
94. B
95. C
96. C
97. C
98. C
99. C
100. A

Phase II, Exam II: Rationale & References for Questions

Question #1. The fire service is typically a team organization in every aspect. NFPA 4.2.2, 5.2. *CO, 2E,* Page 125.

Question #2. ICS should be used during any and all incidents—training or real. NFPA 4.6, 4.7. *CO, 2E,* Page 308.

Question #3. Each worker's compensation claim is money paid out. NFPA 5.4.2, 5.7.1. *CO, 2E,* Page 177.

Question #4. Most of the time, officers deal with other functions rather than emergency response. NFPA 1021 4.2. *CO, 2E,* Page 5.

Question #5. The initial action plan dictates the outcome based upon strategies set and actions or tactics deployed. NFPA 4.6, 5.6. *CO, 2E,* Page 320.

Question #6. It is important to be able to manage time to have good efficiency. NFPA 4.7, 5.7.1. *CO, 2E,* Page 105.

Question #7. Most structured departments have a pyramid-style design that it is important to be able to communicate from top to bottom and bottom to top. NFPA 4.2. *CO, 2E,* Page 60.

Question #8. Taking a look at the potentials is the beginning stages to developing action plans. NFPA 4.6.1, 5.6.1. *CO, 2E,* Page 292.

Question #9. This is a cycle due to the achievements and progressional growth of personnel. NFPA 5.2.2. *CO, 2E,* Page 152.

Question #10. The officer helps the firefighter develop based on the community needs. NFPA 2-2.1. *CO, 2E,* Page 13.

Question #11. Arson has to be targeted at someone. NFPA 4.5.1, 4.5.2, 5.5.2. *CO, 2E,* Page 271.

Question #12. Most fires are not investigated by an investigator, but by the company officer. NFPA 4.5.1, 4.5.2, 5.5.2. *CO, 2E,* Page 276.

Question #13. Planning is a systematic process and should be on going beginning with the end in mind. NFPA 5.2.1, 4.2.6, 4.2.1. *CO, 2E,* Page 99.

Question #14. Most officers have high power needs. NFPA 5.2. *CO, 2E,* Page 125.

Question #15. This is due to the fact that most firefighters die early on in incidents and hose advancement is generally in early stages of incidents. NFPA 4.7.2, 5.7.1. *CO, 2E,* Page 178.

Question #16. Most officers enjoy the challenges with which they are faced because, to get to the officer level, you usually are a high achiever. NFPA 5.2 , 4.2.2. *CO, 2E,* Page 125.

Question #17. Taking a look at the potentials is the beginning stages to developing pre-incident plans. NFPA 4.6.1, 5.6. *CO, 2E,* Page 299.

Question #18. The fire officer helps meet all of the objectives and the grading of ISO. NFPA 5.1.1. *CO, 2E,* Page 86.

Question #19. Lightweight construction is in Type II or Type V based on if it is wood or steel. NFPA 4.6.1. *CO, 2E,* Page 251.

Question #20. Being able to write is an important part of your professional career. NFPA 4.1.2. *CO, 2E,* Page 28.

Question #21. We all like a good working environment. As a company officer, you should help ensure that the hygiene factors are there. This will increase productivity and morale. NFPA 5.7.1, 4.7.1. *CO, 2E,* Page 124 & 125.

Question #22. Inspections are a proactive approach to fire prevention and a great customer service NFPA 5.5.1. *CO, 2E,* Page 226.

Question #23. Arson is an area that must be proven in a court of law. Not every illegally set fire is arson. NFPA 4.5.1. *CO, 2E,* Page 222.

Question #24. As company officers, you have authority and you have the power to use it. NFPA 4.2. *CO, 2E,* Page 62.

Question #25. An officer with line authority manages one or more of the functions that are essential for the fire department's mission. When we see an organizational chart, we usually think of the authority one has and in the areas they have it. NFPA 4.4.2; 5.2.1. *CO, 2E,* Page 62.

Question #26. This is an EEOC directive. NFPA 5.2.1; 4.2.1. *CO, 2E,* Page 133.

Question #27. We as officers must take responsibility for our actions to be responsible for the actions of others. NFPA 4.2.6 section A. *CO, 2E,* Page 62.

Question #28. The mission statement is the road map for the destination of the organization. NFPA 4.2. *CO, 2E,* Page 87.

Question #29. We have this due to the uniform, badge, and title. NFPA 4.2, 5.2. *CO, 2E,* Page 127.

Question #30. Managing time is an important aspect of a company officer. Time is the one thing that is hard to track for efficiency based upon the tasks performed and the personnel performing them. It is important, however, not to fall into this concept of allowing the prescribed time to dictate the time needed for the event. NFPA5.2.1; 5.2.2; 5.2.3; 4.4.2. *CO, 2E,* Page 1058.

Question #31. The company officer is responsible for many items. Operations account for 10 percent and administrative aspects 90 percent whereas company officers do reports, evaluations, and other administrative document record keeping. NFPA 4.2.1, 4.2.2. *CO, 2E,* Page 5.

Question #32. Building construction and fire behavior go hand in hand. It is important to understand how fire will behave in particular build construction styles. NFPA 4.6.1. *CO, 2E,* Page 248 & 249.

Question #33. Reporting to one boss is the concept in this portion. Unity of command is an essential organizational concept. NFPA 4.2.2; 5.2.1. *CO, 2E,* Page 66.

Question #34. Critical incident stress is a significant part of an emergency responder's life. The need for assistance is often necessary. NFPA 4.6.4. *CO, 2E,* Page 206.

Question #35. This is much like a private counseling session to set direction for the subordinate. NFPA 5.2.2. *CO, 2E,* Page 181.

Question #36. We are given this legitimate power when we are promoted to the officer level by the governing body. NFPA 4.2, 5.2. *CO, 2E,* Page 127.

Question #37. Officers should exercise active listening to make sure they are getting the message. It builds respect with the employees. NFPA 4.2.2 section A. *CO, 2E,* Page 24.

Question #38. As officers, it is important to understand the causes of fires and fire fatalities so you can reinforce the materials delivered during fire safety programs and contacts. NFPA 4.5.1. *CO, 2E,* Page 222.

Question #39. An important part of fire prevention and the company officer's job is to enforce codes that are applicable to the jurisdiction. Many of these cross between building and fire prevention. It is important to understand how each relates. NFPA 5.5.1. *CO, 2E,* Page 228.

Question #40. This is the act of guiding the human and physical resources of an organization to attain the organization's objectives. The company officer is the key in this principle whereas they are the ones managing the work force. NFPA4.2.1; 5.2.1. *CO, 2E,* Page 75.

Question #41. Policies give us direction to follow to remain consistent. NFPA 4.2.5. *CO, 2E,* Page 84.

Question #42. NFPA 4.6.2, 5.6.2. *CO, 2E,* Page 257.

Question #43. Company officers are the first level of supervisor and this is where most employees look to for leadership. NFPA 4.2.1. *CO, 2E,* Page 62.

Question #44. This is a standard set out by NFPA 1500. NFPA 5.7, 4.7. *CO, 2E,* Page 184.

Question #45. Rehabilitation is a significant requirement to assure that the personnel working on an emergency scene are physically able to continue working. This is a way to monitor health conditions and potentially recognize signs that could indicate when firefighters are not medically well to continue working. This concept could potentially lower the firefighter death rate in the heart attack area significantly. NFPA 4.7.1. *CO, 2E,* Page 197.

Question #46. The sender must formulate a thought and send it out to a receiver. NFPA 4.1.2. *CO, 2E,* Page 21.

Question #47. Action plan puts the planning and thinking phases into motion. NFPA 4.6 , 5.6. *CO, 2E,* Page 334.

Question #48. The medium is important because this is where the message could be lost or not understood due to barriers. NFPA 4.1.2. *CO, 2E,* Page 21.

Question #49. When acquired structures are used for training, they must be carefully inspected and prepared for the training evolution. This standard provides those guidelines. NFPA 4.7.1. *CO, 2E,* Page 310.

Question #50. Good communications is essential in both work and personal life. NFPA 4.1.2. *CO, 2E,* Page 21.

Question #51. NFPA 4.7. *CO, 2E,* Page 214.

Question #52. When the strategy is well defined, all personnel understand the tasks at hand and can focus their efforts on making it happen. NFPA 4.6, 5.6. *CO, 2E,* Page 333.

Question #53. Working in a union environment is a very different situation and will vary from union to union. It is important to understand how the union functions and its goals. NFPA 4.2, 5.2.1 5.1.1. *CO, 2E,* Page 102.

Question #54. As company officers, we affect the budget in our everyday activities. NFPA 4.4.3, 5.4.2. *CO, 2E,* Page 103.

Question #55. NFPA 4.6.1, 5.6.1. *CO, 2E,* Page 254.

Question #56. This standard provides the basic requirements for firefighters in their proficiency at the job requirements. NFPA 4.7.1. *CO, 2E,* Page 308.

Question #57. To have our firefighter population meet the needs of the citizens, it is important to have the total community in view. This means to have representation of various cultures working to help create better understanding and service needs to the entire community. NFPA 4.7.1. *CO, 2E,* Page 132.

Question #58. Fire investigators must work carefully to locate, gather, and preserve evidence. NFPA 4.5.1, 4.5.2, 5.5.2. *CO, 2E,* Page 281.

Question #59. An organizational chart defines the roles and lines of authority that are important in a paramilitary organizational structure. It is important to know where you fall in the organizational chart as a company officer. NFPA 4.4; 4.6; 5.4; 5.6. *CO, 2E,* Page 47 & 48.

Question #60. Physical fitness lowers injury and death rates by conditioning firefighters for the job. It is important as an officer to know what standards are in place for safety. NFPA 4.7.1; 4.7.2. *CO, 2E,* Page 202.

Question #61. This is where individual employees are equal in abilities and tasks. The span of control must be followed. NFPA 4.4.2. *CO, 2E,* Page 67 & 68.

Question #62. We work in diverse communities and it is essential to have a diverse workforce. NFPA 4.2.1, 5.2.1. *CO, 2E,* Page 132.

Question #63. Most officers are aware of the operating budget since this is the concept we use most in our personal lives. NFPA 4.4.3, 5.4.2. *CO, 2E,* Page 103.

Question #64. The fire department is a group of people working to accomplish a task. NFPA 4.4.2, 5.2.1. *CO, 2E,* Page 119.

Question #65. Spoken word is the most common used method of communication. NFPA 4.1.2. *CO, 2E,* Page 22.

Question #66. Leadership is accomplishing the organization's goals through others. NFPA 4.4.1. *CO, 2E,* Page 120.

Question #67. You will be required to take disciplinary actions as an officer. It is important to use them as a tool for redirection of personnel onto the department's mission pathway. NFPA 5.2.1. *CO, 2E,* Page160.

Question #68. Coaching is an informational process that helps subordinates improve their skills and abilities. Coaching implies one-on-one relationship that treats subordinates as full partners. NFPA 5.2.1. *CO, 2E,* Page 147.

Question #69. NFPA 4.6.2, 5.6.1. *CO, 2E,* Page 323 - 326.

Question #70. Harassment of any kind should not be allowed in the workplace. Title VII of the Civil Rights Act of 1964 NFPA 4.7.1. *CO, 2E,* Page 133.

Question #71. This is a phase that you will go through as you transition from one phase to another. You would not start out in this mode. NFPA 4.6, 5.6. *CO, 2E,* Page 329.

Question #72. It is important to understand the growth and progression of fire to aid in developing action plans. NFPA 4.6.2. *CO, 2E,* Page257 - 258.

Question #73. In active listening, you can actually show the sender that you are actively listening by focusing all of your attention on the speaker and showing a genuine interest in the message. Active listening is a significant portion of good communications. NFPA 5.2 : 4.2. *CO, 2E,* Page 24.

Question #74. Human behavior is important in the leadership and management styles you choose as an officer. NFPA 4.2.4, 5.2.1. *CO, 2E,* Page 123.

Question #75. Operations personnel are called line personnel whereas they refer to the analogy of military as we are engaged in battles with the responses we encounter. NFPA 4.2.6. *CO, 2E,* Page 63 & 64.

Question #76. Controlling helps us get to the right place at the right time through the monitoring of efforts of the resources. This is a large part of the company officer's job. NFPA 4.2.1. *CO, 2E,* Page 79.

Question #77. It is important to understand managerial theories to understand why personnel act and do as they do. It helps in developing leadership styles. NFPA 4.2.6 , 4.4.2. *CO, 2E,* Page 82.

Question #78. The most valuable resource in the fire service is personnel. NFPA 4.2.2. *CO, 2E,* Page 54.

Question #79. We do a lot of this leadership style as employees begin to come into their own talents. NFPA 4.2.3, 5.2.2. *CO, 2E,* Page147.

Question #80. This is the professional standard to which fire officers are trained and certified. NFPA 1021. *CO, 2E,* Page 9.

Question #81. Management can be enhanced through good policy, procedures, and even by personal observation. Policies are broad in nature and should be clearly understood. NFPA 4.3.1; 5.2.1. *CO, 2E,* Page 84.

Question #82. Certification means that an individual has been tested by an accredited examining body on clearly identified material and found to meet the minimum standard. NFPA 3.3.11. *CO, 2E,* Page 8.

Question #83. There are many types of organizations. The important part is that it is a group of people working to accomplish a specific set of goals or functions. NFPA 4.2. *CO, 2E,* Page 119.

Question #84. Many organizations and groups will work together on scenes. It is important to have one leader who is supported by a great staff whereas the leader can make decisions and communicate them. NFPA 4.6.2 section B. *CO, 2E,* Page 66.

Question #85. With rescue being the number one priority in emergency responses, it is important to know pre-designated locations where individuals will go to reach safety during an emergency. NFPA 4.6.3 5.6.1. *CO, 2E,* Page 298.

Question #86. It is important to understand the fire problem in America. NFPA 5.5, 4.7. *CO, 2E,* Page 214.

Question #87. We climb the ladder of rank in the fire service progressing upward. NFPA 4.4. *CO, 2E,* Page 66.

Question #88. Standards set the lower limits for any business. NFPA 3.3.3. *CO, 2E,* Page 9.

Question #89. To make a gear system run, it must be synchronized. The controlling of personnel efforts is like a machine running with the employees being the gears. NFPA 4.4.2, 4.6.2. *CO, 2E,* Page 78.

Question #90. Company officers are the first supervisors in the pyramid structure organization. NFPA N/A. *CO, 2E,* Page 4.

Question #91. Having an overview of exposures and the other hazards are easily identified in this style of preplan. NFPA 4.6.1. *CO, 2E,* Page 295.

Question #92. To accomplish the goals of the organization, it is important to utilize all of your resources. NFPA 4.6.3. *CO, 2E,* Page 67.

Question #93. Barriers can obscure messages. NFPA 4.2.1 section A. *CO, 2E,* Page 23 & 24.

Question #94. As a company officer, you have to envision what is in the future and plan for it. NFPA 4.4. *CO, 2E,* Page 77.

Question #95. It is important to understand the causes of injuries and death of firefighters to take a proactive approach to reduce these numbers. NFPA 4.7.1; 4.7.2. *CO, 2E,* Page 179.

Question #96. The command sequence is a three-step process that helps the incident commander effectively manage the scene. NFPA 4.6 , 5.6. *CO, 2E,* Page 322.

Question #97. The building code and fire code go hand in hand. This is a way that a proactive approach can be taken in the developmental stages of a construction project to enhance fire safety. NFPA 5.5. *CO, 2E,* Page 228.

Question #98. Being a role model means being professional. It is important to have all of the necessary qualities to be an effective leader. NFPA 4.2; 5.2. *CO, 2E,* Page 15.

Question #99. This is a proactive approach; it involves the activities that help keep fires from occurring. NFPA 5.5. *CO, 2E,* Page 214.

Question #100. Stairwells and other penetrations to allow for rescue, fire spread, and potential falling hazards for firefighters. These are important to company officers as they work and direct crews in tactical operations. NFPA 4.6.2. *CO, 2E,* Page 248.

Exam III

1. A public fire department is part of which of the following?
   a. community
   b. fire district
   c. organizational development
   d. local government

2. You and your company are the dispatched unit that command has designated as the rapid intervention team. Which of the following best describes your duties per the textbook?
   a. You are the team of firefighters that will have a minimum of two personnel. You are assigned the primary task of rescuing firefighters should the need arise. You will also be immediately available for any task that command has to assign to you.
   b. You are the team of firefighters that will have a minimum of two personnel. You are assigned the primary task of rescuing firefighters should the need arise. You will not be immediately available for any task that command assigns to you.

3. Which of the following is not one of the four career development roles of a leader?
   a. coach
   b. apprentice
   c. adviser
   d. referral agent

4. You are a company officer in a department that has multiple stations. Your station is a multiple-company station and you are a seasoned company officer on a rescue company. Your battalion chief asks you to develop a five year station work plan. One year into this plan, you find that you are behind by two months. During the next year, you make up this deficit and are on schedule as planned. This is an example of _______________.
   a. controlling
   b. commanding
   c. organizing
   d. developing

5. A volunteer fire department would be which of the following?
   a. formal organization
   b. informal organization
   c. combination organization
   d. all of the above

6. Place in order the following pieces of the command sequence: 1. size-up 2. implementing an action plan 3. developing an action plan
   a. 1, 3, 2
   b. 3, 2, 1
   c. 2, 3, 1
   d. 1,2, 3

7. You respond to the older portion of the city for a reported fire in a four-story commercial building. This portion of the city was mostly constructed in the 1920s. Which type of building construction would you anticipate to see?
   a. fire-resistive
   b. non-combustible
   c. heavy timber
   d. ordinary

8. The strategically combined various maneuvers that are used to achieve a strategy while fighting a fire or dealing with a similar emergency is known as ______________.
   a. tactics
   b. strategy
   c. size-up
   d. action plan

9. Your preplans program is underway and assignments have been made. You are instructed to use a plot plan for your drawing on the pre-incident survey sheet. Which of the following best describes the drawing portion?
   a. bird's-eye view of the property
   b. floor plan
   c. engineering schematic
   d. three-dimensional plan

10. Communications that usually have legal standing within the organization are known as ______________.
   a. informal
   b. written
   c. informational
   d. formal

11. William Ouchi developed the management style in which the manager believes that people not only like to work and can be trusted, but that they want to be collectively involved in the management process and recognized when successful. This is known as which of the following theories?
   a. Theory A
   b. Theory X
   c. Theory Z
   d. Theory Y

12. Incident priorities include which of the following?
    a. life safety
    b. incident stabilization
    c. property conservation
    d. all of the above
13. Company officers have status and power. We add this concept to our list of motivators that get employees to do what is asked of them. You ask a firefighter to take out the trash. The task is quickly completed. One reason would be that you as the company officer have ______ power.
    a. legitimate
    b. reward
    c. punishment
    d. identification
14. You are the senior company officer and the first arriving company officer to the scene of a multi-family residential apartment complex on Halloween night. You arrive to find this two-story apartment complex with heavy fire involvement on the first and second floors of one end involving at least four apartments. Dispatch had a subject on the phone begging for help when the phone was dropped and the line remained open for a period of time before going dead. You have a victim unaccounted for at this time. On arrival, what are your sequential incident priorities based upon response to this scenario?
    a. life safety, incident stabilization, property conservation
    b. incident stabilization, property conservation, life safety
    c. life safety, property conservation, incident stabilization
    d. a or b
    e. all of the above
15. During emergency scene operations, the firefighter typically would be the _______ in the communications model.
    a. sender
    b. message
    c. receiver
    d. all of the above
16. Big words, technical terms, and language differences are examples of _______________.
    a. mediums
    b. barriers
    c. understanding
    d. problems

17. Use the Iowa State formula. You respond to a commercial structure which is 150 feet wide, 400 feet long, and 10 feet high on each of 50 floors. With this knowledge, figure fire flow based on an appropriate theoretical fire flow formula utilizing all of the buildings dimensions. How much would the fire flow be (in gpms) if the building was 25% involved on the 8th floor?
    a. 10, 000 gpm
    b. 1,500 gpm
    c. 3,000 gpm
    d. 30,000 gpm
18. An administrative process whereby an employee is punished for not conforming to the organizational rule and regulations is known as _______________.
    a. acceptable
    b. commendation
    c. disciplinary action
    d. all of the above
19. ABC fire department has a progression or career development path for individuals to move up through the ranks from firefighter to chief. This uninterrupted series of steps or layers is an organizational concept known as _______________.
    a. flat principle
    b. scaler principle
    c. unity of command
    d. division of labor
20. A team or company of emergency personnel kept immediately available for the potential rescue of other emergency responders is known as the _______________ team.
    a. safety
    b. rapid response
    c. rapid intervention
    d. rapid safety
21. America Burning was published in _______________.
    a. 1903
    b. 1943
    c. 1973
    d. 1993
22. A graphic representation of what the organization should look like that defines the formal lines of authority and responsibility is known as a(n) _______________.
    a. organizational chart
    b. organizational growth
    c. informal organization
    d. group chart

23. As a newly promoted company officer, you are _______________.
    a. responsible for the performance of the assigned crew
    b. responsible for the safety of the assigned crew
    c. second-line supervisor
    d. a & b
    e. all of the above

24. Which of the following is not a building construction type?
    a. ordinary
    b. heavy roof
    c. non-combustible
    d. heavy timber

25. The act of controlling the fire to prevent the fire from extending any further in the structure is known as _______________.
    a. fire detection
    b. overhaul
    c. extinguishment
    d. confinement

26. A company officer certified as a Fire Officer I should meet the National Fire Protection Association Standard 1021 ________.
    a. Chapter 2
    b. Chapter 3
    c. Chapter 4
    d. Chapter 5

27. A mental assessment of the situation; gathering and analyzing information that is crucial to the outcome of the event is _______________.
    a. pre-planning
    b. pre-incident analysis
    c. size-up
    d. none of the above

28. You are the company officer and an employee comes to you with a problem. They state they have a problem, but also have a potential solution. To assure you have received the complete message, you should practice which of the following?
    a. communications
    b. barrier breakdown
    c. active listening
    d. passive listening

29. You are senior company officer who has been tasked at looking into the health and safety program of the department. You are tasked with bringing the department's rules, regulations, and standard operating guidelines into compliance with professional standards and laws. Which two organizations would you look to for information and guidance?
    a. JEMS and FIREHOUSE magazines
    b. NIOSH and JOCCA
    c. NFPA and JOCCA
    d. NFPA and OSHA

30. Activities that take about as long as the time is allowed is known as a concept called the ______________.
    a. Managerial Law
    b. Parkinson's Law
    c. Range Law
    d. Empowering Law

31. As a company officer, you have assigned personnel. These personnel are dependent upon you for guidance daily. This makes your primary role ______________.
    a. leadership
    b. coaching
    c. mentoring
    d. friendship

32. You are reviewing an employee's performance evaluation for the quarter. It is important to have good communications occurring. This two-way communications allows the employee to provide ___ on the evaluation.
    a. feedback
    b. knowledge
    c. barriers
    d. all of the above

33. If the deputy chief of operations would come by a station to have a friendly lunch with the station personnel, this would be an example of a(n) ______________.
    a. informal group
    b. formal group
    c. formal relationship
    d. informal relationship

34. The difference between the fire triangle and the fire tetrahedron is ______________.
    a. heat
    b. fuel
    c. chemical chain reaction
    d. oxygen

35. You have an employee who brings in a pornographic magazine to work. He shows a picture of a female in it to several members of the crew. He states the female in the picture looks like another department employee. What answer best describes the condition described here?
    a. harassment
    b. sexual harassment
    c. Civil Rights Act violation
    d. b & c
    e. all of the above
36. This sets broad goals and outlines the overall plan to control the incident.
    a. tactics
    b. strategy
    c. size-up
    d. action plan
37. The company officer spends _____ percent of their time dealing with administrative aspects
    a. 10
    b. 40
    c. 50
    d. 90
38. As a fire continues to build, the smoke levels drop. When hot, fuel-rich gases meet fresh air, they ignite and burn along the ceiling. This is known as ______________.
    a. backdraft
    b. flashover
    c. phase II
    d. rollover
39. Firefighting operations that make a direct attack on the fire for purposes of control and extinguishment is _______________ mode.
    a. defensive
    b. transitional
    c. command
    d. offensive
40. Locations where there are unusual hazards, or where an incident would likely overload the department's resources or where there is a need for interagency cooperation to mitigate the hazard is known as _______________.
    a. hazardous material sites
    b. universities
    c. target hazards
    d. none of the above

41. From 2002, the leading cause of death to firefighters was _______________.
    a. responding/returning from alarms
    b. training
    c. fire ground
    d. non-fire emergencies

42. The span of control is important as a company officer is responsible for human resources. According to the Company Officer textbook, the number of personnel that a supervisor can effectively manage is _____ to ______.
    a. 2, 4
    b. 4, 7
    c. 3, 5
    d. 5, 8

43. A characteristic or organizational structure denoting the relationship between supervisors and subordinates is called _______________.
    a. accountability
    b. line authority
    c. responsibility
    d. flow chart

44. You are the senior company officer and the first arriving company officer to the scene of a multi-family residential apartment complex on Halloween night. You arrive to find this two-story apartment complex with heavy fire involvement on the first and second floors of one end involving at least four apartments. Dispatch had a subject on the phone begging for help when the phone was dropped and the line remained open for a period of time before going dead. You now have brought the fire under control. You have a victim unaccounted for at this time. Which of the following meet the guidelines for calling a fire investigator?
    a. potentially incendiary fire
    b. a fire resulting in a death
    c. property damage greater than $50,000
    d. a & b
    e. all of the above

45. The company officer is the crew leader and first-line supervisor of assigned staffing and equipment. We refer to these assigned personnel as _______________.
    a. staff
    b. bodies
    c. human resources
    d. facility resources

46. You are the investigating officer on a small room and contents fire in a bedroom. The fire began about 11 PM on Friday, August 21. You have conducted interviews with the occupants who are smokers and are now looking at findings in the room. The bed is burned significantly and the heat line of demarcation is about 5 feet off the floor. There is heavy smoke residue in the room with minor fire damage to any other furniture.You find a space heater plugged up close to the window beside the bed, several extension cords, and an ash tray by the bed. What would you suspect the cause of this fire to be?
    a. electrical
    b. improperly positioned space heater
    c. smoking
    d. not enough information to determine
47. An obstacle in communications that prevents the message from being understood by the receiver is ______________.
    a. a barrier
    b. reading
    c. active listening
    d. none of the above
48. Which of the following is the leading cause of injuries in fires?
    a. matches
    b. heating equipment
    c. cooking
    d. natural causes
49. You have been a company officer for five years and have been found to be reliable and dependable to the higher administrative staff on smaller assigned projects. The chief has asked you to head up a hiring process for new recruits. It is summer time and you have a new boat you and your family are enjoying everyday you are off. You also have taken several days off to go to the lake. You were assigned the hiring project on June 9. The completion date has been set for July 23. Today it is July 13 and you have done very little work on this project. Which one of the following best describes you in the proven time eaters?
    a. lack of personal goals
    b. reacting to urgent events
    c. procrastination
    d. trying to do too much yourself
50. You have an employee who you feel doesn't like work and you have found that he cannot be trusted to keep deadlines or meet minimum productivity levels. You are describing which of the following management theories?
    a. TQM
    b. Theory X
    c. Theory Y
    d. Theory Z

51. As a company officer, you are issued a legal document that sets forth the requirements for life safety and property protection in the event of fire, explosion, or similar emergency. ITS PURPOSE IS TO MINIMIZE THE RISK OF LOSS OF LIFE AND PROPERTY BY REGULATING THE USE AND STORAGE OF MATERIALS THAT MIGHT BE ON THE PROPERTY. This legal document is which of the following?
    a. Fire Prevention Code
    b. Building Code
    c. Minimal Standards
    d. Elements and Standards

52. The second stage of fire is known as _______________ phase.
    a. incipient
    b. free-burning
    c. backdraft
    d. smoldering

53. You are the senior company officer and the first arriving company officer to the scene of a multi-family residential apartment complex on Halloween night. You arrive to find this two-story apartment complex with heavy fire involvement on the first and second floors of one end involving at least four apartments. Dispatch had a subject on the phone begging for help when the phone was dropped and the line remained open for a period of time before going dead. You have a victim unaccounted for at this time. You are making an interior attack. This would be known as _______________ mode.
    a. offensive
    b. defensive
    c. transitional
    d. none of the above

54. The father of professional management was _______________.
    a. McGregor
    b. Brunacini
    c. Drucker
    d. Fayol

55. The definition of ________ is being responsible for one's personal activities; in the organizational context, this includes being responsible for the actions of one's subordinates.
    a. responsibility
    b. accountability
    c. line authority
    d. management

56. Self-operating thermosensitive device that releases a spray of water over a designated area to control or extinguish a fire is known as _______________.
    a. deluge gun
    b. stand pipe
    c. pre-piped system
    d. automatic fire protection sprinkler

57. The area that tends to cause problems in communications is often the _______________.
    a. sender
    b. message
    c. receiver
    d. medium

58. Which of the following is the first step in the management process involving looking to the future and determining objectives?
    a. organizing
    b. planning
    c. commanding
    d. coordinating

59. You are a company officer who arrives on the scene of a strip mall complex. You see heavy volumes of fire coming through the roof of an anchor store. As which type of building construction would this strip mall be classified?
    a. fire-resistive
    b. non-combustible
    c. heavy timber
    d. frame

60. A financial plan to purchase high-dollar items that have a life expectancy of more than one year is referred to as the _______________ budget.
    a. operating
    b. line-item
    c. capital
    d. program

61. A dramatic event in a room fire that rapidly leads to full involvement of all combustible materials present almost simultaneously is known as _______________.
    a. backdraft
    b. rollover
    c. flashpoint
    d. flashover

62. As the company officer, you will use many leadership styles to accomplish the mission of the department and the goals of the company. You have an employee who is a five-year veteran and has demonstrated progressing strength in skills and abilities on the job. Which leadership style will you most utilize?
    a. directing
    b. consulting
    c. supporting
    d. delegating
63. Which of the following is not a role of the company officer?
    a. coach
    b. counselor
    c. role model
    d. administrator
    e. none of the above
64. Which of the following is a part of effective writing skills?
    a. consider the reader
    b. emphasis
    c. brevity
    d. simplicity
    e. all of the above
65. The final step in the management process involves monitoring the process to ensure that the work is accomplishing the intended goals and objectives and taking corrective action when it is not. This is known as _______________.
    a. coordinating
    b. organizing
    c. controlling
    d. commanding
66. Safety is whose concern?
    a. company officer
    b. everyone
    c. safety officer
    d. fire chief
    e. all of the above
67. A formal document indicating the focus, direction, and values of an organization is _______________.
    a. vision
    b. mission statement
    c. SOPs
    d. rules and regulations

68. Which of the following are proven time eaters?
    a. lack of personal goals and objectives
    b. good planning
    c. procrastination
    d. a & c
    e. b & c

69. A company officer is responsible for the _________ of the assigned personnel in an emergency services organization.
    a. performance
    b. safety
    c. mission development
    d. both a & b
    e. both a & c

70. A document that attests that one has demonstrated the knowledge and skills necessary to function in a particular craft or trade is known as _______________.
    a. qualification
    b. training
    c. certification
    d. all of the above

71. A financial plan to acquire the goods and services needed to run an organization for a specific period of time is referred to as _______________ budget.
    a. line-item
    b. program
    c. capital
    d. operating

72. Collecting similar items into a single account and presenting them in a general area collectively under this topic or one line in a budget document is a(n) _______________ budget.
    a. capital
    b. operating
    c. program
    d. line-item

73. As the company officer, you will use many leadership styles to accomplish the mission of the department and the goals of the company. You have an employee who is a five year veteran and has demonstrated strength in skills and abilities on the job. This employee functions at the next level consistently and routinely takes on projects and requires little support from you. Which leadership style will you most utilize?
    a. directing
    b. consulting
    c. supporting
    d. delegating

74. The organizational concept that refers to the uninterrupted series of steps or layers within an organization refers to which of the following?
    a. linear principle
    b. unity principle
    c. scaler principle
    d. division of labor principle

75. As an applicant with a department, your employment is conditional on your joining the union. Failure to join or agreement to join resulting in your employment is an example of a(n) _______ shop.
    a. open
    b. closed
    c. union
    d. agency

76. As a company officer, in which of the following three essential elements in fire prevention would you be the most involved?
    a. education
    b. engineering
    c. enforcement

77. The accomplishment of the organization's goals by utilizing the resources available is done through the leadership of _______________.
    a. management
    b. command
    c. unity of direction
    d. Fayol

78. You are a company officer in a department that has multiple stations. Your station is a multiple-company station and you are a seasoned company officer on a rescue company. Your battalion chief asks you to develop a five-year station work plan. What type of plan would you be developing?
    a. long-range term
    b. mid-range plan
    c. short- range plan
    d. mini plan

79. You have an employee who has just gone through the death of their spouse, leaving them with two children and a large sum of bills due to a long illness. This employee has spent countless hours caring for their spouse and continuing to work and care for the children. Which of the following best describes the position in which this person would fall in Maslow's hierarchy of needs?
    a. self-actualization
    b. self-esteem
    c. security needs
    d. social needs

80. A company in a fire department is which of the following?
    a. formal organization
    b. informal organization
    c. formal organization within an informal organization
    d. all of the above

81. The typical downtown business districts in older communities where the buildings are generally no more than five stories tall and the exterior is masonry are generally known as which type of building construction?
    a. Type I
    b. Type II
    c. Type III
    d. Type IV

82. The critical process of shifting from the offensive mode to the defensive mode or the defensive mode to the offensive mode is known as _______________ mode.
    a. switching
    b. moving
    c. transitional
    d. flip flop

83. Communications that are more simple and spontaneous are known as _______________.
    a. formal
    b. informal
    c. written
    d. none of the above

84. Which of the following is a legal document that sets forth requirements to protect health, safety, and the general welfare of the public as they relate to construction and occupancy of buildings and construction?
    a. fire prevention code
    b. building code
    c. city codes
    d. fire codes

85. You respond to a reported structure fire in a university dormitory where there are known invalids. The plan for them is to protect in place or go to a portion of the structure that is relatively safe from fire and the byproducts of fire. This area is known as an area of __________.
    a. security
    b. safety
    c. congregating
    d. refuge

86. The deliberate and apparent process by which one focuses attention on the communications of another is known as _______________.
    a. listening
    b. active listening
    c. passive listening
    d. aggressive listening

87. You are a newly promoted lieutenant in your department. Where do you fit into the department's organization?
    a. middle management
    b. lower management
    c. first-line supervisor
    d. a & c
    e. b & c

88. The difference between the fire triangle and the fire tetrahedron is _______________.
    a. heat
    b. fuel
    c. chemical chain reaction
    d. oxygen

89. An organizational principle that addresses the number of personnel a supervisor can effectively manage is the _______________.
    a. command
    b. unity of command
    c. delegation
    d. span of control

90. Significant points in the emergency event usually marking the accomplishments of one of three incident priorities. This is known as _______________.
    a. benchmarking
    b. timemarking
    c. trendmarking
    d. time elapsed

91. You are the company officer who arrives on scene of a 30-story building with fire showing from one window on the 22nd floor. The building described here will most likely be made of which type of construction?
    a. fire-resistive
    b. non-combustible
    c. woodframe
    d. ordinary

92. Under the needs theory, a firefighter who you supervise takes responsibility for their efforts and has far exceeded the training requirements for two positions above their position. They have set even higher goals with you to continue to grow. You would say this person is which of the following?

    a. high-achievement needs

    b. high-affiliation needs

    c. low-achievement needs

    d. low-affiliation needs

93. You are assigned as the leader of a task force that will be working in a collapsed building. You have selected team members that are specifically trained in the area of structural collapse with specialties in shoring, engineering design, and fiber optic visual technology. As the task force leader you would be considered to be___________ this group as you begin rescue operations for trapped victims.

    a. organizing

    b. commanding

    c. developing

    d. coordinating

94. You are a company officer in a department that has multiple stations. Your station is a multiple company station and you are a newly promoted company officer on a rescue company. Your battalion chief asks you to develop a work plan for the next year. What type of plan would you be developing?

    a. long-range plan

    b. mid-range plan

    c. short-range plan

    d. mini plan

95. A person who helps a team member or subordinate improve their knowledge, skills, and abilities is known as a _______________.

    a. counselor

    b. coach

    c. director

    d. chief

96. You as the company officer have been instructed by the fire chief to write a letter to a citizen referencing a complaint about the use of sirens during the middle of the night. Which of the following is important?

    a. technical terms

    b. emphasis

    c. considering who the reader is

    d. complexity

97. Company officers are expected to be _______ to their assigned personnel.
    a. full time instructors
    b. full-time leaders
    c. both a & c
    d. none of the above

98. Use the Iowa State formula. You respond to a commercial structure which is 150 feet wide, 500 feet long, and 12 feet high on each of 50 floors. With this knowledge, figure fire flow based on an appropriate theoretical fire flow formula utilizing all of the building's dimensions. How much would the fire flow be (in gpms) if the building was 25% involved on the 50th floor?
    a. 10, 000 gpm
    b. 2,250 gpm
    c. 3,000 gpm
    d. 30,000 gpm

99. As a company officer, you are faced with a question for which you are not sure of the answer. What should you do?
    a. Turn it back to the employee to find the answer.
    b. Admit you don't know the answer.
    c. Try to find the answer.
    d. a & b
    e. b & c

100. As a company officer, you are always looking for the advantage over emergency responses especially fire. Which one of the following would be classified as one of our greatest allies in fire protection?
    a. construction features
    b. fire codes
    c. sprinkler systems
    d. smoke detectors

Phase II, Exam III: Answers to Questions

1.	D	26.	C	51.	A	76.	A
2.	B	27.	C	52.	B	77.	A
3.	B	28.	C	53.	A	78.	B
4.	A	29.	D	54.	D	79.	C
5.	A	30.	B	55.	B	80.	A
6.	A	31.	A	56.	D	81.	C
7.	D	32.	A	57.	D	82.	C
8.	D	33.	D	58.	B	83.	B
9.	A	34.	C	59.	B	84.	B
10.	D	35.	E	60.	C	85.	D
11.	C	36.	B	61.	D	86.	B
12.	D	37.	D	62.	B	87.	E
13.	C	38.	D	63.	D	88.	C
14.	A	39.	D	64.	A	89.	D
15.	C	40.	C	65.	C	90.	A
16.	B	41.	C	66.	E	91.	A
17.	B	42.	B	67.	A	92.	A
18.	C	43.	B	68.	D	93.	B
19.	B	44.	E	69.	D	94.	C
20.	C	45.	C	70.	C	95.	B
21.	C	46.	C	71.	D	96.	C
22.	A	47.	A	72.	D	97.	C
23.	D	48.	C	73.	D	98.	B
24.	B	49.	C	74.	C	99.	E
25.	D	50.	B	75.	B	100.	C

Phase II, Exam III: Rationale & References for Questions

Question #1. It is important to understand the infrastructure of the community and government of which the fire department is a part. NFPA 4.3.1. *CO, 2E,* Page 54.

Question #2. Each year, over 100 firefighters die in the line of duty. The majority of the fire ground deaths occur when firefighters are advancing hose lines inside the structure. NFPA 4.7.1. *CO, 2E,* Page 197.

Question #3. As a leader, you will be in one of these roles with different subordinates at the same time. It is important to understand as the individual grows, your role will change. NFPA 5.2.1. *CO, 2E,* Page 151.

Question #4. Controlling allows us to measure the effectiveness of our effort to help us maintain our goals. By doing so, we can seek ways to improve, thus increasing productivity. NFPA 5.6.1; 5.4.2. *CO, 2E,* Page 79.

Question #5. Groups exist whenever two or more people share a common goal. Organizations are groups of people. Typically, these people share a common goal, have formal rules, and have designated leaders. This is true of fire companies. NFPA 4.2. *CO, 2E,* Page 44-46.

Question #6. The command sequence is a three-step process that helps incident commanders manage the incident. NFPA 4.6.2. *CO, 2E,* Page 323-334.

Question #7. This style of construction is referred to as Main Street, USA. NFPA 4.6.2. *CO, 2E,* Page 250.

Question #8. The action plan addresses all phases of the incident. NFPA 4.6, 5.6. *CO, 2E,* Page 331.

Question #9. These are utilized for primary access points, barriers to access, utilities, and water supply. They work well for a quick 20-second reference. NFPA 4.6.1. *CO, 2E,* Page 295.

Question #10. Good communication skills are essential in work and personal life. Formal communications are conducted according to established standards. They tend to follow customs, rules, and practices. NFPA 4.2.2; 5.2.1. *CO, 2E,* Page 20.

Question #11. As a company officer, it is important to understand which style of management practices you must choose. Each situation will require a different style or use of theory. NFPA 5.2.1. *CO, 2E,* Page 81.

Question #12. Strategy and tactics are set to accomplish the goals for the incident. The priorities are set so as to provide a guide to what areas and tactics need to be taken care of in a sequential order if the situation dictates. NFPA 4.6, 5.6. *CO, 2E,* Page 320.

Question #13. The recognition of the power of position accompanied by the authority by virtue of the supervisor's ability to administer punishment leads others to see the punishment power concept. NFPA 5.2.1. *CO, 2E,* Page127.

Question #14. The priorities for any incident are as follows, in order: life safety, incident stabilization, and property conservation. NFPA 4.6.2. *CO, 2E,* Page 320.

Question #15. As the company officer gives orders on the emergency scene, the firefighter, who is the receiver of the message, interprets the message and acts from there. NFPA 4.2.2. *CO, 2E,* Page 21.

Question #16. There are many barriers to effective communications. These are some of the common ones. NFPA 5.2; 4.2. *CO, 2E,* Page 24.

Question #17. Length x Width x Height / 100 = the fire flow for each floor. If the floor is 1/4 involved, then divide that answer by 4 and you would get 1,500 gpm. NFPA 4.6.1. *CO, 2E,* Page 264.

Question #18. Disciplinary action is another way of improving performance and skills. This lets the subordinate know that their performance is below acceptable standards. NFPA 5.2.2. *CO, 2E,* Page 160.

Question #19. Principle of scaler organizations is that as you progress with experience and talents you move upward to higher levels. This is important whereas the company officer is the first level on which individuals accept responsibility for themselves and others. NFPA 4.2. *CO, 2E,* Page 66.

Question #20. This concept is indicated in NFPA 1500 where emphasis on accountability and safety during emergency operations has significant emphasis. NFPA 4.7. *CO, 2E,* Page 197.

Question #21. NFPA 4.7. *CO, 2E,* Page 214.

Question #22. An organizational chart defines the roles and lines of authority that are important in a paramilitary organizational structure. It is important to know where you fall in the organizational chart as a company officer. NFPA 4.4; 4.6; 5.4; 5.6. *CO, 2E,* Page 47-48.

Question #23. The company officer's job is as varied as all the activities that the organizations does. However, the company officer is the first-line supervisor and generally the senior representative with which the public deals on a routine basis. NFPA 4.1.1. *CO, 2E,* Page 4.

Question #24. Building construction will define fire behavior and the tactics required to extinguish a fire. NFPA 4.6.2. *CO, 2E,* Page 248-251.

Question #25. Benchmarks or progression in fire attack is important to the company officer as definitive positions in the incident. NFPA 4.6.3. *CO, 2E,* Page 335.

Question #26. It is important to understand the requirements to which you are certifying as a company officer. NFPA 4.1. *CO, 2E,* Page 9-11.

Question #27. Remember that size-up is a good first step in developing an action plan. NFPA 4.6.2, 5.6.1. *CO, 2E,* Page 323-326.

Question #28. The deliberate and apparent process by which one focuses attention on the communications of another is active listening. This is important to the company officer so as to get the full message of the individual presenting the problem and solution. NFPA 4.2.2. *CO, 2E,* Page 24.

Question #29. Safety comes from the elimination of unsafe acts or equipment. It is important as a company officer to recognize the issues and then the requirements and standards that support those issues. NFPA 4.7.1. *CO, 2E,* Page 182-183.

Question #30. Managing time is an important aspect of a company officer. Time is the one thing that is hard to track for efficiency based upon the tasks performed and the personnel performing them. It is important, however, not to fall into this concept of allowing the prescribed time to dictate the time needed for the event. NFPA 5.2.1; 5.2.2; 5.2.3; 4.4.2. *CO, 2E,* Page 105.

Question #31. Leading others is the company officer's primary job as you are the one who is assigned to direct the personnel assigned to you. To accomplish the mission, you must lead others. NFPA 4.2.1. *CO, 2E,* Page 5.

Question #32. As a company officer, it is important to remember to allow in this situation the communications to be a two-way process. NFPA 4.2.2. *CO, 2E,* Page 21.

Question #33. Although the fire service is a formal organization, informal relationships exist as well just as described in this question where a superior officer has an informal lunch with the crew of a station. NFPA 4.2.1; 4.2.2; 4.2.3. *CO, 2E,* Page 60.

Question #34. Understanding fire growth is important as a company officer as you make multiple split-second decisions on tactics and operations based upon the knowledge of fire behavior. NFPA 4.6.2. *CO, 2E,* Page 257.

Question #35. Supervisors should be sure that all employees are fairly treated and represented in all department activities. NFPA 5.2.1; 4.7.1; 5.2.2. *CO, 2E,* Page 131.

Question #36. Strategies are fundamental to planning and directing operations. NFPA 4.6, 5.6. *CO, 2E,* Page 333.

Question #37. The company officer is responsible for many items. Operations relate for 10 percent and administrative aspects 90 percent whereas we do reports, evaluations, and other administrative document record keeping. NFPA 4.2.1, 4.2.2. *CO, 2E,* Page 5.

Question #38. It is important to understand the growth and progression of fire in developing action plans. NFPA 4.6.2. *CO, 2E,* Page 258.

Question #39. During an offensive mode, operations are conducted in an aggressive manner. NFPA 4.6 5.6. *CO, 2E,* Page 327.

Question #40. Target hazards are those occupancies that present high risk to life safety and property. Given that you cannot plan for every situation, you should plan for those that present the greatest risk. NFPA 4.6.2, 5.6.1. *CO, 2E,* Page 293.

Question #41. It is important to understand the causes of injuries and death of firefighters to take a proactive approach to reduce these numbers. NFPA 4.7.1; 4.7.2. *CO, 2E,* Page 179.

Question #42. As a company officer, it is important to be able to manage incidents and daily functions. The use of span of control is important to be able to effectively manage these personnel. NFPA 4.2. *CO, 2E,* Page 67.

Question #43. An officer with line authority manages one or more of the functions that are essential for the fire department's mission. When we see an organizational chart, we usually think of the authority one has and in the areas they have it. NFPA 4.4.2; 5.2.1. *CO, 2E,* Page 62.

Question #44. The experience of a fire investigator will prove to be beneficial especially with the elements of potential death and potential arson. NFPA 4.5.1; 4.5.2; 5.5.2. *CO, 2E,* Page 277.

Question #45. Whether volunteer or career, the company officer is responsible for the human resources or staffing along with other resources. NFPA 4.2.1; 4.2.2; 4.2.3. *CO, 2E,* Page 54.

Question #46. Knowing the leading causes of fire are being careless smoking and the evidence presented, you would tend to lean toward that answer based upon fire behavior and the time of day coupled with the ashtray finding and the smoking resident. NFPA 4.5.1. *CO, 2E,* Page 221.

Question #47. Consider a barrier to be like a filter. One or more filters reduces the information flow between the sender and the receiver. NFPA 4.2; 5.2. *CO, 2E,* Pages 23.

Question #48. Understanding the nation's fire problem will help you focus your goal-setting toward helping make change. NFPA 4.5.1, 5.5.2. *CO, 2E,* Page 222.

Question #49. The inability to say no or decline the opportunity is the hardest of the time eaters for most. You must remember to tie yourself to goals and focus on priorities. It is important to know when you are overloading your plate. Procrastination is just the opposite. You do not have a focus on goals or priorities as it relates to work. The continued putting off of this project has you in a position of crunch time for completion. This usually leads to poor work. NFPA 5.2.1. *CO, 2E,* Page 106.

Question #50. This theory is a management theory introduced by Douglas McGregor. This management style in which the manager believes that people dislike work and cannot be trusted was a traditional style of management. NFPA 5.2.1; 5.2.2. *CO, 2E,* Page 81.

Question #51. This is the document by which you inspect. These codes are adopted by local, state, or other jurisdictional entities. NFPA 5.5.1. *CO, 2E,* Page 228.

Question #52. Flames become visible along with an increased amount of heat and light generation. NFPA 4.6.2 , 5.6.2. *CO, 2E,* Page 257.

Question #53. The priorities for any incident are as follows in order: life safety, incident stabilization, and property conservation. Operations that make a direct attack on the fire are in the offensive strategy to control the fire and keep it in place. NFPA 4.6.3. *CO, 2E,* Page 327.

Question #54. Fayol is known as the father of management. He is known for his research and development of managerial principles. NFPA 5.2.1. *CO, 2E,* Page 77.

Question #55. This is the other responsibility as you accept responsibility for others and their actions as a company officer. NFPA 4.2. *CO, 2E,* Page 62.

Question #56. As a fire officer, you should understand these systems and how to operate them. NFPA 4.6.1, 5.6.1. *CO, 2E,* Page 253.

Question #57. Mediums can be a variety of problems to include noise, language, terminology, and others. It is important to try to control this area as much as possible so your communications will be as effective and as clear as possible. NFPA 4.3.3; 4.2.1; 5.2.1; 5.2.2. *CO, 2E,* Pages 21.

Question #58. Planning covers everything from the next hour to the next decade. NFPA 4.2.3. *CO, 2E,* Page 77.

Question #59. Because this type of construction allows long spans without support, it is very popular in large one-story commercial facilities. NFPA 4.6.2. *CO, 2E,* Page 249.

Question #60. Classic examples of capital budget items are fire stations, apparatus purchase, and any large ticket item. It is important to understand the budgeting process whereas you are managing the money at the first level in protecting the organization's investments. NFPA 4.4.3; 5.4.2. *CO, 2E,* Page 103.

Question #61. Flashover is a fire phenomena that presents significant risk to firefighters. Fire behavior is a crucial link to how we formulate tactics to accomplish our strategic goals. NFPA 4.7.1, 4.6.2, 5.6.2. *CO, 2E,* Page 258.

Question #62. Most employees described in the question require some discussion, direction, and supervision. They are beginning to contribute ideas and solutions. NFPA 5.2.1, 4.2.1. *CO, 2E,* Page 128.

Question #63. The roles of company officers are many. It is important to know your roles as an officer. Effective company officers fill many roles throughout their careers. Many are done simultaneously. NFPA 4.2. *CO, 2E,* Page 5.

Question #64. The importance of written communications is imperative. Many of these documents will be official records and will also serve as information being sent to the public. NFPA 4.3.3; 5.2.2; 5.2.1. *CO, 2E,* Page 28.

Question #65. Controlling helps us get to the right place at the right time through the monitoring of efforts of the resources. This is a large part of the company officer's job. NFPA 4.2.1. *CO, 2E,* Page 79.

Question #66. Safety comes from the elimination of unsafe equipment and procedures, having and enforcing policies that provide for safety, and having viable recognition programs at all levels that reinforce both the company's and industry's commitment to worker safety and health. NFPA 4.7.1; 4.7.2; 5.7.1. *CO, 2E,* Page 178-179.

Question #67. The mission statement declares the vision of the department by setting specific values and setting the focus on which direction the department is moving. A mission statement is like putting up a sign for employees on which direction to go. NFPA 5.2; 4.2. *CO, 2E,* Page 97-98.

Question #68. People allow important events to become urgent events through the lack of good time management techniques. Putting off work creates rushes and inefficiency. This precludes lower quality in work performance, thus results in hurting the organization. NFPA 4.4.2; 5.2.1. *CO, 2E,* Page 106.

Question #69. The role of the company officer is to manage personnel at the crew level. Performance and safety are key items for which the officer is responsible. NFPA 4.2.1. *CO, 2E,* Page 4.

Question #70. Certification means that an individual has been tested by an accredited examining body on clearly identified material and found to meet the minimum standard. NFPA 3.3.11. *CO, 2E,* Page 8.

Question #71. This is everything not covered typically in the capital budget. The operating budget is a general budget that shows specific amounts needed to operate the organization in general form. NFPA 5.4.2; 4.4.3. *CO, 2E,* Page 103.

Question #72. Line-item budgeting is like keeping accounting on the money spent. This is used when you need specifics to be purchased and don't want it to be generic in nature. NFPA 4.4.3; 5.4.2. *CO, 2E,* Page 103.

Question #73. Most employees described in the question require little discussion, direction,and supervision. They are at the next level waiting for challenges and opportunities to present themselves. NFPA 5.2.1, 4.2.1. *CO, 2E,* Page 128.

Question #74. Like playing a scale on a musical instrument where every note is sounded, the scaler principle suggests that every level in the organization is considered in the flow of communications. NFPA 4.4.2. *CO, 2E,* Page 66.

Question #75. As an organization that is a bargaining union, it is important for an officer to understand the factors that surround the contract and the department so as not to violate any laws or conditions that are legally binding. It is important also to know these as you are the representative for your employees' best welfare. NFPA 5.2.1. *CO, 2E,* Page 102.

Question #76. This area is generally left to fire protection engineers and building code officials to assure they meet standards and will protect the structure. NFPA 5.5.1; 4.3.4. *CO, 2E,* Page 223.

Question #77. This is the act of guiding the human and physical resources of an organization to attain the organization's objectives. The company officer is the key in this principle whereas they are the ones managing the work force. NFPA 4.2.1; 5.2.1. *CO, 2E,* Page 75.

Question #78. The first step in the management process is planning and the effectiveness of a company officer has the ability to make and keep plans moving forward. NFPA 5.4.5. *CO, 2E,* Page 77.

Question #79. A good leader must understand human behavior and the needs of employees as the life outside of the workplace often influences life in the work atmosphere. NFPA 4.7; 5.7. *CO, 2E,* Page 123.

Question #80. Groups exist whenever two or more people share a common goal. Organizations are groups of people. Typically, these people share a common goal, have formal rules, and have designated leaders. This is true of fire companies. NFPA 4.2. *CO, 2E,* Page 44-46.

Question #81. Action plans for fire attack are based upon fire behavior and thc building construction to determine the appropriate tactics. NFPA 4.6.1. *CO, 2E,* Page 249.

Question #82. This is a mode with which you will never begin, rather a phase that you pass through as you change modes along with strategies and tactics. NFPA 4.6 5.6. *CO, 2E,* Page 329.

Question #83. We write with less formality when sending emails, memos, and short notes. We also tend to be less formal at social events. NFPA 5.2.1; 4.2.1,; 4.2.2. *CO, 2E,* Page 20.

Question #84. An important part of fire prevention and the company officer's job is to enforce codes that are applicable to the jurisdiction. Many of these cross between building and fire prevention codes. It is important to understand how each relates. NFPA 5.5.1. *CO, 2E,* Page 234.

Question #85. . *CO, 2E,* Page 298.

Question #86. In active listening, the listener can actually show the sender that you are actively listening by focusing all of your attention on the speaker and showing a genuine interest in the message. Active listening is a significant portion of good communications. NFPA 5.2 : 4.2. *CO, 2E,* Page 24.

Question #87. Company officers, for firefighters, are the first step in the chain of command. Command-level officers consider the company officer to be the lower management staff and the first-line supervisors for the organization. NFPA 4.2. *CO, 2E,* Page 4.

Question #88. Understanding fire growth is important as a company officer as you make multiple split-second decisions on tactics and operations based upon the knowledge of fire behavior. NFPA 4.6.2. *CO, 2E,* Page 272.

Question #89. The number of people you can effectively supervise varies, of course, based on many factors, but generally for our purposes, the number is between four and seven. NFPA 5.2.1; 5.6.1. *CO, 2E,* Page 67.

Question #90. As an incident commander, it is important to understand the work progress made during an incident. Benchmarking helps the incident commander keep track of tasks completed and the location of crew members for personnel accountability. NFPA 4.6, 4.7, 5.6, 5.7. *CO, 2E,* Page 320.

Question #91. All high rise buildings are made of fire-resistive construction. NFPA 4.6.2. *CO, 2E,* Page 248.

Question #92. Individuals in theory with high achievement needs tend to work more diligently and want to see it as an achievement. NFPA 5.2.2. *CO, 2E,* Page 125.

Question #93. The third step in the management process, commanding, involves using the talents of others, giving them directions, and setting them to work. NFPA 5.6.1. *CO, 2E,* Page 78.

Question #94. The first step in the management process is planning. The effectiveness of a company officer is the ability to make and keep plans moving forward. NFPA 5.4.5. *CO, 2E,* Page 77.

Question #95. Coaching is an informational process that helps subordinates improve their skills and abilities. Coaching implies one-on-one relationship that treats subordinates as full partners. NFPA 5.2.1. *CO, 2E,* Page 147.

Question #96. In writing, it is important to remember who is going to read what you wrote. This way, you will write to have effective communications with no barriers. NFPA 4.3.2. *CO, 2E,* Page 28.

Question #97. These two pieces are some of the most important parts of what company officers do in a daily function. Human relations are extremely important to the crews and the success of the company officer. NFPA 4.2.2 ; 4.2.3; 5.2.1; 5.2.2. *CO, 2E,* Page 5-7.

Question #98. Length x Width x Height / 100 = the fire flow for each floor. If the floor is 1/4 involved, then divide that answer by 4 and you would get 2,250 gpm. NFPA 4.6.1. *CO, 2E,* Page 264.

Question #99. In the communications process you will be faced with questions to which you may not know the answer. Be up front and honest that you don't know the answer, but work to find the answer. In doing so, you will enhance your knowledge and develop good communications skills. NFPA 4.2.6. *CO, 2E,* Page 26-28.

Question #100. Fire suppression systems like sprinklers were designed to control the fire growth or even suppress the fire. With this assistance, we already have no reported deaths within sprinklered buildings. This system is like giving the fire companies additional personnel for suppression efforts that are housed in each occupancy. NFPA 4.6.2; 4.6.3. *CO, 2E,* Page 253.

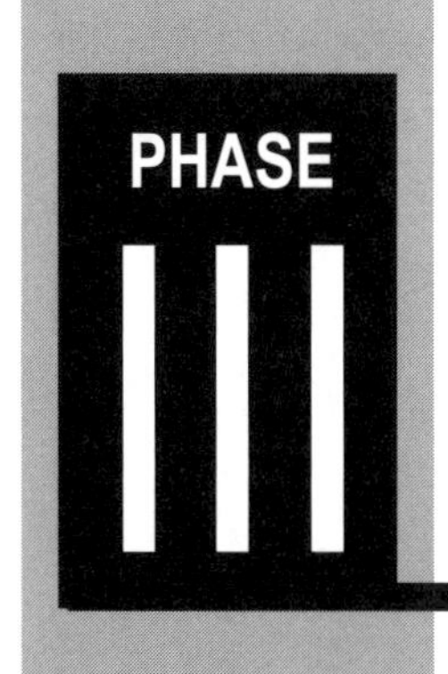

SYNTHESIS & EVALUATION

Referring to Table I-1, the final levels of Bloom's Taxonomy, Cognitive Domain, are covered in section three. Mastery of this section suggests the highest level of understanding of the material. The levels addressed are:

- synthesis
- evaluation

The successful test-taker should be able to rearrange material, modify processes, compare data, and interpret results. For testing at this level, questions will be tied around more of an application process. The student should be able to apply the information learned in the class or in the textbook.

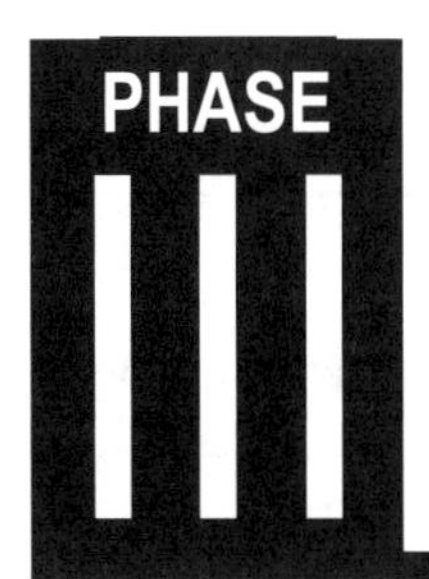

Exam I

1. You respond to an automatic alarm and are the second apparatus to arrive. Your standard assignment is to take the water supply and the Fire Department Connection. You know this based on the procedures that are written known as which of the following?
   a. quick access pre-plans
   b. notes
   c. standard operating procedures
   d. none of the above

2. As a company officer, upon promotion you are afforded the right and power to command. This right and power by position is known as _______________.
   a. respect
   b. power
   c. managerial legal right
   d. authority

3. As a company officer, in which of the following three essential elements in fire prevention would you be the least involved?
   a. education
   b. engineering
   c. enforcement

4. You are assigned as the leader of a task force that will be working in a collapsed building. You have selected team members that are specifically trained in the area of structural collapse with specialties in shoring, engineering design, and fiber optic visual technology. As the task force leader, you would be considered to be___________ this group as you begin rescue operations for trapped victims.
   a. organizing
   b. commanding
   c. developing
   d. coordinating

5. As the company officer, you will use many leadership styles to accomplish the mission of the department and the goals of the company. You have an employee who is fresh out of the academy. Which leadership style will you most likely utilize?
   a. directing
   b. consulting
   c. supporting
   d. delegating

6. Certification means that an individual has been tested by an accredited examining body on clearly identified materials and found to meet the minimal standard. There are other reasons to certify. Which of the following is not one of the reasons to certify?

   a. protection from liability
   b. recognition of demonstrated proficiency
   c. recognition of professionalism
   d. all of the above
   e. b & c only

7. You are the company officer of an employee who has thrown a fire cracker in the day room where there are several other employees. The fire cracker detonated. This caused one of the employees to have some pain in their ear. You are going to take disciplinary action on the employee's behavior. Which of the following would be the most appropriate disciplinary form?

   a. transfer
   b. demotion
   c. oral reprimand
   d. termination

8. If the deputy chief of operations would come by a station to have a friendly lunch with the station personnel this would be an example of a(n) _______________.

   a. informal group
   b. formal group
   c. formal relationship
   d. informal relationship

9. You respond to a residential structure fire that has been vacant and for rent over the last year. The electrical meter base is capped with no meter in place. The fire started on the second floor. The rear door is open and you find strange patterns of char on the floor with close areas not burned. What is your assumption on the cause of this fire?

   a. electrical cord under the rugs
   b. arson
   c. electrical wiring fire in the floor
   d. none of the above

10. Apparatus and activities are ongoing on the emergency scene. These pieces are considered to be which of the following in the communications process?

    a. message
    b. barriers
    c. feedback
    d. transmitter

11. You are the company officer determining an employee's competency and commitment. This employee has strong skills and self-confidence. They are always looking for ways to get out of work or avoid projects on which to work. This employee would be which of the following?
    a. high competency / high commitment
    b. low competency / high commitment
    c. high competency / low commitment
    d. low competency / low commitment
12. You are a newly promoted company officer. From where will most of your management occur?
    a. behind a desk
    b. in the field
    c. in the fire station
    d. none of the above
13. For what does the "C" in "COAL WAS WEALTH" stand?
    a. cold
    b. construction
    c. controlled
    d. clear
14. Your company has been assigned to a city festival in the Fire Department's booth with life safety educators. Your interaction and primary responsibilities for the assignment would be _______________.
    a. tactical
    b. standing by
    c. fire suppression
    d. fire prevention
15. You are the company officer determining an employee's competency and commitment. This employee has strong skills and self-confidence. They are always looking for ways to get involved or projects on which to work. This employee would be which of the following?
    a. high competency / high commitment
    b. low competency / high commitment
    c. high competency / low commitment
    d. low competency / low commitment
16. As a company officer, you may be assigned to manage the department's health and safety program. If you had this assignment, you would have the responsibility and title of _____.
    a. safety officer
    b. health and safety
    c. health and safety officer
    d. none of the above

17. As a company officer, you have the ability to get many tasks accomplished. When promoted, you were automatically given power through the position. By this power you make employees work to levels to which they may not normally be self-motivated to work. This concept is known as________ power.
    a. reward
    b. legitimate
    c. punishment
    d. persuasion

18. You respond to the older portion of the city for a reported fire in a four-story commercial building. This portion of the city was mostly constructed in the 1920s. Which type of building construction would you anticipate to see?
    a. fire-resistive
    b. non-combustible
    c. heavy timber
    d. ordinary

19. You are the company officer and an employee comes to you with a problem. They state they have a problem, but also have a potential solution. To assure you have received the complete message, you should practice which of the following?
    a. communications
    b. barrier breakdown
    c. active listening
    d. passive listening

20. You are the company officer determining an employee's competency and commitment. This employee has weak skills and self-confidence. They are always looking for ways to get out of work or avoid projects on which to work. This employee would be which of the following?
    a. high competency / high commitment
    b. low competency / high commitment
    c. high competency / low commitment
    d. low competency / low commitment

21. You arrive on scene of a two-story woodframe single residential structure that you were dispatched to for an automatic fire alarm. Upon entering the structure you find a light haze of smoke on the first floor. Further investigation leads you to a smoldering waste basket in the bathroom. This fire was in which of the following stages?
    a. incipient
    b. free-burning
    c. backdraft
    d. smoldering

22. You respond to a commercial structure which is 150 feet wide, 400 feet long, and 10 feet high. With this knowledge, you could figure fire flow based on which theoretical fire flow formula?
    a. National Fire Academy Formula
    b. Iowa State University Formula
    c. West Virginia University Formula
    d. a & b
    e. a & c

23. As the company officer, you will use many leadership styles to accomplish the mission of the department and the goals of the company. You have an employee who is a five year veteran and has demonstrated strength in skills and abilities on the job. This employee functions at the next level consistently and routinely takes on projects and requires little support from you. Which leadership style will you most likely utilize?
    a. directing
    b. consulting
    c. supporting
    d. delegating

24. If you are preplanning an old cotton mill, you would anticipate the construction to be _______________.
    a. ordinary
    b. Type III
    c. heavy timber
    d. Type V
    e. none of the above

25. You are the shift training officer and you have been given a house in which to do live fire training. You will be following NFPA 1403. Your structure is a single-story house with 2,210 square feet. It has a large fireplace in the living room and a basement with narrow steps leading down off of the kitchen. What will be required to make this structure safe?
    a. floors, railing, and stairs made safe
    b. asbestos removed
    c. chimney hazards eliminated
    d. extraordinary dead weight removed
    e. all of the above

26. As a company officer, you are delegated a project to develop a public education prop for fire prevention week. You are to make a prop that resembles a fire truck for a puppet show. You utilize one of your crew member's talent in carpentry. This concept is known as _______________.
    a. delegation
    b. advocation
    c. management
    d. supervision

27. The span of control is important as a company officer is responsible for human resources. According to the Company Officer textbook, the number of personnel that a supervisor can effectively manage is _____ to ______.
    a. 2, 4
    b. 4, 7
    c. 3, 5
    d. 5, 8

28. You have left the scene of a fire. You are presented with an anonymous tip of information not found on the initial investigation. Which of the following are ways the fire department may reenter the property after they have left the scene?
    a. administrative search warrant
    b. administrative process
    c. owner's consent
    d. search warrant
    e. all of the above

29. As a company officer, you will set levels of achievement with your employees that can be effectively measured to assure that achievement has occurred. In setting these levels, you are helping the employee set obtainable ______.
    a. objectives
    b. JPRs
    c. progress
    d. goals

30. You are on the second alarm responding to a large building fire. Knowing you will be the first engine arriving on the scene in the second alarm, the incident commander summons you to the command post and assigns your crew to work as part of a task force with another company. You are given the task of tracking personnel and their location during the incident. This process is known as _______________.
    a. incident command
    b. unified command
    c. personnel accountability
    d. personnel management

31. You and your company are the dispatched unit that command has designated as the rapid intervention team. Which of the following best describes your duties per the textbook.
    a. You are the team of firefighters which will have a minimum of two personnel. You are assigned the primary task of rescuing firefighters should the need arise. You will also be immediately available for any task that command has to assign to you.
    b. You are the team of firefighters which will have a minimum of two personnel. You are assigned the primary task of rescuing firefighters should the need arise. You will not be immediately available for any task that command has to assign to you.

32. As an applicant with a department, your employment is conditional on your joining the union. Failure to join or agreement to join resulting in your employment is an example of a(n) _______ shop.
    a. open
    b. closed
    c. union
    d. agency

33. The term company is used to describe work teams. The individual of that work team who must coordinate activities is the company officer. This individual has many roles to fill. Which of the following is not one of those roles?
    a. help in advancing hose
    b. directing crew operations
    c. activities that don't compromise roles as leaders
    d. developing a mission statement

34. A volunteer fire department would be which of the following?
    a. formal organization
    b. informal organization
    c. combination organization
    d. all of the above

35. Company officers have status and power. We add this concept to our list of motivators that get employees to do what is asked of them. You ask a firefighter to take out the trash. The task is quickly completed. One reason would be that you, as the company officer, have ______ power.
    a. legitimate
    b. reward
    c. punishment
    d. identification

36. You, as the company officer, have been instructed by the fire chief to write a letter to a citizen referencing a complaint about the use of sirens during the middle of the night. Which of the following is important?
    a. technical terms
    b. emphasis
    c. considering the reader
    d. complexity

37. As a company officer, you have assigned personnel. These personnel are dependent upon you for guidance daily. This makes your primary role ______________.
    a. leadership
    b. coaching
    c. mentoring
    d. friendship

38. You are the investigating officer on a small room and contents fire in a bedroom. The fire began about 11 PM on Friday August 21. You have conducted interviews with the occupants who are smokers and are now looking at findings in the room. The bed is burned significantly and the heat line of demarcation is about 5 feet off the floor. There is heavy smoke residue in the room with minor fire damage to any other furniture.You find a space heater plugged in close to the window beside the bed, several extension cords, and an ashtray by the bed. What would you suspect the cause of this fire to be?
    a. electrical
    b. improperly positioned space heater
    c. smoking
    d. not enough information to determine

39. You are a company officer in a department that has multiple stations. Your station is a multiple-company station and you are a seasoned company officer on a rescue company. Your battalion chief asks you to develop a five-year station work plan. One year into this plan, you find that you are behind by two months. During the next year, you make up this deficit and are on schedule as planned. This is an example of ______________.
    a. controlling
    b. commanding
    c. organizing
    d. developing

40. As a company officer, you must look at the needs for your crew and develop them accordingly. The first step in this management process would be ______________.
    a. planning
    b. delegating
    c. centralizing
    d. advocating

41. As a company officer, you are always looking for the advantage over emergency responses especially fire. Which one of the following would be classified as one of our greatest allies in fire protection?
    a. construction features
    b. fire codes
    c. sprinkler systems
    d. smoke detectors

42. You are a company officer in a department that has multiple stations. Your station is a multiple-company station and you are a newly promoted company officer on a rescue company. Your battalion chief asks you to develop a work plan for the next year. What type of plan would you be developing?
    a. long-range term
    b. mid-range plan
    c. short- range plan
    d. mini plan

43. You are the company officer of an employee who has thrown a fire cracker in the day room where there are several other employees. The fire cracker detonated. This caused one of the employees to have some pain in their ear. Which of the following would be the most appropriate answer?

    a. applauding the employee
    b. coaching the employee
    c. counseling the employee
    d. ignoring the situation since it was a joke

44. You are the senior company officer and the first arriving company officer to the scene of a multi-family residential apartment complex on Halloween night. You arrive to find this two-story apartment complex with heavy fire involvement on the first and second floors of one end involving at least four apartments. Dispatch had a subject on the phone begging for help when the phone was dropped and the line remained open for a period of time before going dead. You now have brought the fire under control. You have a victim unaccounted for at this time. Which of the following meets the guidelines for calling a fire investigator?

    a. potentially incendiary fire
    b. a fire resulting in a death
    c. property damage greater than $50,000
    d. a & b
    e. all of the above

45. As the company officer, you will use many leadership styles to accomplish the mission of the department and the goals of the company. You have an employee who is a five year veteran and has demonstrated progressing strength in skills and abilities on the job. Which leadership style will you most likely utilize?

    a. directing
    b. consulting
    c. supporting
    d. delegating

46. A company in a fire department is which of the following?

    a. formal organization
    b. informal organization
    c. formal organization within an informal organization
    d. all of the above

47. Career development is a shared responsibility between which of the following?

    a. the individual and the community
    b. the individual and the department
    c. the community and the department
    d. the environment and you

48. You are the company officer and one of your employees feels that your disciplinary actions are too harsh for the incident. You and the employee are unable to come to an agreement even with formal meetings. You would expect the employee to start a ________________.

    a. gripe
    b. grievance
    c. conflict
    d. complaint

49. The addition of a sprinkler system to a building lowers which of the following?

    a. life risk factors
    b. property risk factors
    c. physical risk factors
    d. community risk factors

50. You have an employee who you feel doesn't like work and you have found that he cannot be trusted to keep deadlines or meet minimum productivity levels. You are describing which of the following management theories?

    a. TQM
    b. Theory X
    c. Theory Y
    d. Theory Z

51. As a company officer, you will open channels of communications between employees and other department ranks with your crew members. In doing so, you often persuade the two sides to meet to discuss and work out differences. This concept is known as ________________.

    a. meditation
    b. arbitration
    c. mediation
    d. addiction

52. A company officer certify as a Fire Officer I should meet the National Fire Protection Association Standard 1021 _________.

    a. Chapter 2
    b. Chapter 3
    c. Chapter 4
    d. Chapter 5

53. You are a company officer who is in charge of a crew who is stationed in a mixed district representing some business and residential response potential. In the response district, the residential configurations range from million dollar single-family homes to multi-family housing to mobile homes. This mixture will bring a diverse group of residents that depend on you for services. To understand all of the population, it is important to have a crew strong in which of the following areas?
    a. emergency scene operations
    b. cultural diversity
    c. tactical considerations
    d. all of the above

54. You have been a company officer for five years and have been found to be reliable and dependable to the higher administrative staff on smaller assigned projects. The chief has asked you to head up a hiring process for new recruits. It is summer time and you have a new boat you and your family are enjoying everyday you are off. You also have taken several days off to go to the lake. You were assigned the hiring project on June 9. The completion date has been set for July 23. Today it is July 13 and you have done very little work on this project. Which one of the following best describes you in the proven time eaters?
    a. lack of personal goals
    b. reacting to urgent events
    c. procrastination
    d. trying to do too much yourself

55. The company officer is the crew leader and first line supervisor of assigned staffing and equipment. We refer to these assigned personnel as _______________.
    a. staff
    b. bodies
    c. human resources
    d. facility resources

56. As a company officer, it is necessary to provide a good work environment for your employees. Herzberg's Model describes what is needed to get employees to come to work and not be dissatisfied. Items like good working relationships and considerate supervisors are examples of _________.
    a. hygiene factors
    b. motivators
    c. achievements
    d. power

57. You are the company officer who arrives on scene of a 30-story building with fire showing from one window on the 22nd floor. The building described here will mostly likely be made of which type of construction?
    a. fire-resistive
    b. non-combustible
    c. woodframe
    d. ordinary

58. ABC fire department has a progression or career development path for individuals to move up through the ranks from Firefighter to chief. This uninterrupted series of steps or layers is an organizational concept known as _______________.

    a. linear principle
    b. scaler principle
    c. unity of command
    d. division of labor

59. You have an employee who is working hard and making great progression in their career. Everything in their life seems to be just perfect. According to Maslow this individual would be in which position in his hierarchy of needs model?

    a. social needs
    b. security needs
    c. self-actualization needs
    d. survival mode

60. You are reviewing an employee's performance evaluation for the quarter. It is important to have good communications occurring. This two-way communications allows the employee to provide ___ on the evaluation.

    a. feedback
    b. knowledge
    c. barriers
    d. all of the above

61. As a company officer, you meet with your crew and set a vision for where the company is going based on the department's mission statement. You set clear concise expectations based upon the guiding principles of the department. In developing this, you are focusing on both the internal and external customer's needs. In doing this, you are practicing what management style?

    a. Theory X
    b. Theory Z
    c. traditional management
    d. total quality management

62. Which one of the following has the greatest dollar loss attributed to its damages annually?

    a. hurricanes
    b. fire
    c. tornadoes
    d. floods

63. You have been a company officer for five years and have been found to be reliable and dependable to the higher administrative staff on assigned projects. The more you have proven your talents, the more special assignments you seem to be getting pushed your way. Now you are finding you are struggling to get routine tasks accomplished. Which one of the following best describes you in the proven time eaters?

    a. lack of personal goals
    b. reacting to urgent events
    c. inability to say no
    d. trying to do too much yourself

64. Pre-incident planning consists of which of the following activities?

    a. pre-incident survey
    b. development of information resources that would be useful during the event
    c. development of procedures that would be used during the event
    d. a & c
    e. all of the above

65. You have an employee who has just gone through the death of their spouse leaving them with two children and a large sum of bills due to a long illness. This employee has spent countless hours caring for their spouse and continuing to work and care for the children. Which of the following best describes the position into which this person would fall in Maslow's Hierarchy of needs.

    a. self-actualization
    b. self-esteem
    c. security needs
    d. social needs

66. You are a newly promoted lieutenant in your department. Where do you fit into the department's organization?

    a. middle management
    b. lower management
    c. first-line supervisor
    d. a & c
    e. b & c

67. You are the investigating officer on a small room and contents fire in a bedroom. The fire began about 11 PM on Friday, February 1. The temperature outside is -10 degrees and there is 8 inches of snow on the ground. You have conducted interviews with the occupants who are smokers and are now looking at findings in the room. The bed is burned significantly and the heat line of demarcation is about 5 feet off the floor. There is heavy smoke residue in the room with minor fire damage to any other furniture.You find a space heater plugged in close to the window beside the bed, several extension cords, and an ashtray by the bed. What would you suspect the cause of this fire to be?

    a. electrical
    b. improperly positioned space heater
    c. smoking
    d. not enough information to determine

68. You are inside a working structure fire and you encounter heat that is significant in nature near the 3-foot mark off of the floor. Below that mark, it is significantly cooler. In fire behavior, you know that the heat is an indication of the potentials that exist. We know this concept on higher temperatures closer to the ceiling as_______________.

    a. thermal stratification
    b. line of demarcation
    c. thermal plane
    d. thermal balance

69. As a company officer, you have assigned members of your crew to do preplanning activities. This is a new program the department is trying to get established. You have acquired outside training in the subject matter from an accredited institution. You have spent considerable time training the crew members and have given instructions and made assignments. What is described here is an example of _______________.

    a. organization
    b. coordination
    c. management
    d. leadership

70. You are the senior company officer and the first arriving company officer to the scene of a multi-family residential apartment complex on Halloween night. You arrive to find this two-story apartment complex with heavy fire involvement on the first and second floors of one end involving at least four apartments. Dispatch had a subject on the phone begging for help when the phone was dropped and the line remained open for a period of time before going dead. You have a victim unaccounted for at this time. You are making an interior attack. This would be known as _______________ mode.

    a. offensive
    b. defensive
    c. transitional
    d. none of the above

71. As a company officer, in which of the following three essential elements in fire prevention would you be the most involved?
    a. education
    b. engineering
    c. enforcement

72. The publication that was published in 1973 that included 90 recommendations that could help reduce fire loss in America was________.
    a. American Firefighter
    b. America Burned
    c. America Burning
    d. America Bonding

73. You are the senior company officer and the first arriving company officer to the scene of a multi-family residential apartment complex on Halloween night. You arrive to find this two-story apartment complex with heavy fire involvement on the first and second floors of one end involving at least four apartments. Dispatch had a subject on the phone begging for help when the phone was dropped and the line remained open for a period of time before going dead. You have a victim unaccounted for at this time. On arrival, what are your sequential incident priorities based upon response to this scenario?
    a. life safety, incident stabilization, property conservation
    b. incident stabilization, property conservation, life safety
    c. life safety, property conservation, incident stabilization
    d. a or b
    e. all of the above

74. You respond to a reported structure fire in a university dormitory where there are known invalids. The plan for them is to protect in place or go to a portion of the structure that is relatively safe from fire and the byproducts of fire. This area is known as an area of _______________.
    a. security
    b. safety
    c. congregating
    d. refuge

75. As a company officer effective communications will _______________. 1. enhance leadership ability 2. gain respect from peers and superiors 3. assist in dealing with the media
    a. 1 & 2
    b. 2 & 3
    c. 1, 2, & 3
    d. 1 only
    e. 3 only

76. You arrive on the fire scene and give a size-up. Upon entering the structure, you are advising command you are entering with four personnel. When you begin your attack, you announce water on the fire. These are examples of _______________.
    a. tactics
    b. strategies
    c. flow procedures
    d. benchmarks

77. You are the senior company officer and find yourself in the command role. You have companies that are attacking the fire, addressing exposure, ventilation, water supply, and rescue. With this, you have organized a course of action that addresses all aspects for controlling the incident. This is known as a _______________.
    a. plan of accord
    b. action plan
    c. resource plan
    d. action series

78. As a company officer, which of the following three essential elements in fire prevention would you be the most involved in if you were doing in-service company inspections?
    a. education
    b. engineering
    c. enforcement

79. Old wood frame structure which were built in the early 1900s and late 1800s will probably be built in a style known as _______________.
    a. platform
    b. old school
    c. oak frame
    d. balloon frame

80. Orders given on the emergency scene during operations by the company officer are in the form of _______________.
    a. informal
    b. formal
    c. oral
    d. a & c
    e. b & c

81. You are a company officer who arrives on the scene of a strip mall complex. You see heavy volumes of fire coming through the roof of an anchor store. As which type of building construction would this strip mall be classified?
    a. fire-resistive
    b. non-combustible
    c. heavy timber
    d. frame

82. You have an employee who brings in a pornographic magazine to work. He shows a picture of a female in the magazine to several members of the crew. He states the female in the picture looks like another department employee. What answer best describes the condition described here?
    a. harassment
    b. sexual harassment
    c. Civil Rights Act violation
    d. b & c
    e. all of the above

83. Lightweight scissor wood trusses are commonly used in which of the following types of construction?
    a. Type I
    b. Type II
    c. Type III
    d. Type IV
    e. Type V

84. Your pre-plans program is under way and assignments have been made. You are instructed to use a plot plan for your drawing on the pre-incident survey sheet. Which of the following best describes the drawing portion?
    a. bird's-eye view of the property
    b. floor plan
    c. engineering schematic
    d. three-dimensional plan

85. You have a large retirement facility with skilled nursing facility included. This type of complex is an example of _______________.
    a. fire targets
    b. target hazards
    c. facilities with multiple automatic alarms daily
    d. none of the above

86. As a newly promoted company officer, you are _______________.
    a. responsible for the performance of the assigned crew
    b. responsible for the safety of the assigned crew
    c. the second-line supervisor
    d. a & b
    e. all of the above

87. Place in order the following pieces of the command sequence: 1. size-up 2. implementing an action plan 3. developing an action plan
    a. 1, 3, 2
    b. 3, 2, 1
    c. 2, 3, 1
    d. 1,2, 3

88. Under the needs theory, a firefighter who you supervise takes responsibility for their efforts and has far exceeded the training requirements for two positions above their position. They have set even higher goals with you to continue to grow. You would say this person is which of the following?
    a. high-achievement needs
    b. high-affiliation needs
    c. low-achievement needs
    d. low-affiliation needs

89. The ABC fire department has a pyramid organizational structure. Where does the company officer fall within this structure?
    a. front-line supervisor
    b. first-line supervisor
    c. second-level supervisor
    d. a & b
    e. a & c

90. The definition of ________ is being responsible for one's personal activities; in the organizational context, it includes being responsible for the actions of one's subordinates.
    a. responsibility
    b. accountability
    c. line authority
    d. management

91. A company officer who is conducting an in-service inspection can enforce which of the following?
    a. fire prevention code
    b. electrical code
    c. building code
    d. a & c
    e. a & b

92. Use the Iowa State formula. You respond to a commercial structure which is 150 feet wide, 400 feet long, and 10 feet high on each of 50 floors. With this knowledge, figure fire flow based on an appropriate theoretical fire flow formula utilizing all of the building's dimensions. How much would the fire flow be (in gpms) if the building was 25% involved on the 8th floor?
    a. 10, 000 gpm
    b. 1,500 gpm
    c. 3,000 gpm
    d. 30,000 gpm

93. Use the Iowa State formula. You respond to a commercial structure which is 150 feet wide, 500 feet long, and 12 feet high on each of 50 floors. With this knowledge, figure fire flow based on a appropriate theoretical fire flow formula utilizing all of the building's dimensions. How much would the fire flow be (in gpms) if the building was 25% involved on the 50th floor?
    a. 10, 000 gpm
    b. 2,250 gpm
    c. 3,000 gpm
    d. 30,000 gpm

94. During emergency scene operations, the firefighter typically would be the _______ in the communications model.
    a. sender
    b. message
    c. receiver
    d. all of the above

95. You are a senior company officer who has been tasked at looking into the health and safety program of the department. You are tasked with bringing the department's rules, regulations, and standard operating guidelines into compliance with professional standards and laws. Which two organizations would you look to for information and guidance?
    a. JEMS and FIREHOUSE Magazines
    b. NIOSH and JOCCA
    c. NFPA and JOCCA
    d. NFPA and OSHA

96. You are a company officer in a department that has multiple stations. Your station is a multiple company station and you are a seasoned company officer on a rescue company. Your battalion chief asks you to develop a five-year station work plan. What type of plan would you be developing?
    a. long-range term
    b. mid-range plan
    c. short- range plan
    d. mini plan

97. As a company officer, you are faced with a question and are not sure of the answer. What should you do?
    a. Turn it back to the employee to find the answer.
    b. Admit you don't know the answer.
    c. Try to find the answer.
    d. a & b
    e. b & c

98. As a company officer, you are issued a legal document that sets forth the requirements for life safety and property protection in the event of fire, explosion, or similar emergency. Its purpose is to minimize the risk of loss of life and property by regulating the use and storage of materials that might be on the property. This legal document is which of the following?

   a. fire prevention code
   b. building code
   c. minimal standards
   d. elements and standards

99. Professionalism encompasses attitude, behavior, communication, style, demeanor, and ethical beliefs. Which of the following best describes attitude?

   a. the core of your performance
   b. nothing that helps job performance
   c. you are working to be a good role model
   d. all of the above
   e. a & c

100. As a company officer, you will receive a larger work load and responsibility with the position. Often times, tasks are assigned to the company officer to complete. As work loads increase it is often easy to become overloaded. This is due to the officer not delegating appropriate assignments and trying to do the entire assignment themselves. This occurs when officers will say "I can do it better myself," "It will take too much time to train someone to do that task," or "I really can't take a chance on that project." These are examples of _______________ delegation.

   a. barriers to
   b. reasons for
   c. hemisphere of
   d. facts about

Phase III, Exam I: Answers to Questions

1.	C	26.	C	51.	C	76.	D
2.	D	27.	B	52.	C	77.	B
3.	B	28.	E	53.	B	78.	C
4.	B	29.	D	54.	C	79.	D
5.	A	30.	C	55.	C	80.	E
6.	D	31.	B	56.	A	81.	B
7.	C	32.	B	57.	A	82.	E
8.	D	33.	D	58.	B	83.	E
9.	B	34.	A	59.	C	84.	A
10.	B	35.	C	60.	A	85.	B
11.	C	36.	C	61.	D	86.	D
12.	B	37.	A	62.	B	87.	A
13.	B	38.	C	63.	C	88.	A
14.	D	39.	A	64.	E	89.	D
15.	A	40.	A	65.	C	90.	B
16.	C	41.	C	66.	E	91.	D
17.	B	42.	C	67.	B	92.	B
18.	D	43.	C	68.	A	93.	B
19.	C	44.	E	69.	D	94.	C
20.	D	45.	B	70.	A	95.	D
21.	A	46.	A	71.	A	96.	B
22.	D	47.	B	72.	C	97.	E
23.	D	48.	B	73.	A	98.	A
24.	C	49.	A	74.	D	99.	A
25.	E	50.	B	75.	C	100.	A

Phase III, Exam I: Rationale & References for Questions

Question #1. This is a standardized way of functioning based upon an organized directive that establishes a standard course of action. NFPA 4.6.3; 5.6.1. *CO, 2E,* Page 304.

Question #2. The position of rank affords the right and the power to command. This is the authority you are granted as a company officer. NFPA 4.4.1. *CO, 2E,* Page 99.

Question #3. This area is generally left to fire protection engineers and building code officials to assure they meet standards and will protect the structure. NFPA 5.5.1; 4.3.4. *CO, 2E,* Page 223.

Question #4. The third step in the management process, commanding, involves using the talents of others, giving them directions, and setting them to work. NFPA 5.6.1. *CO, 2E,* Page 78.

Question #5. Most new employees require a lot of direction and supervision. NFPA 5.2.1, 4.2.1. *CO, 2E,* Page 128.

Question #6. Regardless of whether we are paid or volunteer, we are serving the public and want to be considered and viewed as professionals. NFPA 4.1.1. *CO, 2E,* Page 8.

Question #7. The purpose of the disciplinary action is to improve the subordinate's performance or conduct. NFPA 4.2.5. *CO, 2E,* Page 161.

Question #8. Although the fire service is a formal organization, informal relationships exist as well just as described in this question where a superior officer has an informal lunch with the crew of a station. NFPA 4.2.1; 4.2.2; 4.2.3. *CO, 2E,* Page 60.

Question #9. With the situation described as an unoccupied structure without power, there is little chance of an electrical fire. Arson is motivated by spite, fraud, intimidation, and concealment of a crime etc. NFPA 4.5.1. *CO, 2E,* Page 274.

Question #10. There are many barriers or obstacles to overcome on emergency scenes. Noise from these are the ones that exist on the most frequent basis. NFPA 4.2.1, 4.2.2. *CO, 2E,* Page 23.

Question #11. Understanding an employee's competency and commitment is important as you mentor and work with these individuals as a supervisor. NFPA 4.2.2. *CO, 2E,* Page 146.

Question #12. As most fire chiefs would agree they manage primarily from behind a desk, they would also agree that the company officer is a working foreman managing personnel from where the action is occurring just like a floor supervisor in an industrial plant. NFPA 5.6.1; 4.6.3. *CO, 2E,* Page 84.

Question #13. This acronym helps with the size-up process. NFPA 4.6.2. *CO, 2E,* Page 326.

Question #14. Fire prevention is a proactive approach that is most often accomplished through public contact and educational sessions done by firefighters. NFPA 4.3.4. *CO, 2E,* Page 214.

Question #15. Understanding an employee's competency and commitment is important as you mentor and work with these individuals as a supervisor. NFPA 4.2.2. *CO, 2E,* Page 146.

Question #16. According to NFPA 1500, every department should have an individual assigned to the duties of health and safety officer. NFPA 5.7.1. *CO, 2E,* Page 184.

Question #17. Company officers have a status and power. We add this concept to our list of motivators that get employees to do what is asked of them. NFPA 5.2.1. *CO, 2E,* Page 127.

Question #18. This style of construction is referred to as Main Street, USA. NFPA 4.6.2. *CO, 2E,* Page 249.

Question #19. The deliberate and apparent process by which one focuses attention on the communications of another is active listening. This is important to the company officer so as to get the full message of the individual presenting the problem and solution. NFPA 4.2.2. *CO, 2E,* Page 24.

Question #20. Understanding an employee's competency and commitment is important as you mentor and work with these individuals as a supervisor. NFPA 4.2.2. *CO, 2E,* Page 146.

Question #21. The first stage of the fire is limited to the materials originally ignited. NFPA 4.6.2; 5.6.1. *CO, 2E,* Page 257.

Question #22. The amount of water that is required theoretically to suppress the fire can be figured by both of these formulas. NFPA 4.6.1. *CO, 2E,* Page 263 & 264.

Question #23. Most employees described in the question require little discussion, direction, and supervision. They are at the next level waiting for challenges and opportunities to present themselves. NFPA 5.2.1, 4.2.1. *CO, 2E,* Page 128.

Question #24. These structures are called mill structures or heavy timber based on that the majority of these structures that were built, were mills. NFPA 4.6.2. *CO, 2E,* Page 250.

Question #25. All elements of NFPA 1403 and water supply standards must be followed for safety concerns. NFPA 4.7.1. *CO, 2E,* Page 309 & 310.

Question #26. The accomplishment of the organization's goals by utilizing the resources available is management. As a company officer, you manage human resources daily. NFPA 4.2.2. *CO, 2E,* Page 75.

Question #27. As a company officer, it is important to be able to manage incidents and daily functions. The use of span of control is important to be able to effectively manage these personnel. NFPA 4.2. *CO, 2E,* Page 67.

Question #28. Knowing the legal system and the required process is a must if this situation was to occur. NFPA 4.5.1; 4.5.2; 5.5.2. *CO, 2E,* Page 281.

Question #29. A goal is a target or object by which achievement can be measured. As company officers, we set departmental, company, and individual goals for ourselves and others daily. NFPA 4.2.3. *CO, 2E,* Page 98.

Question #30. NFPA 1500 provides policies for managing events to include accountability. NFPA 4.7.1; 4.6.3; 5.6.1. *CO, 2E,* Page 197.

Question #31. Each year, over 100 firefighters die in the line of duty. The majority of the fire ground deaths occur when firefighters are advancing hose lines inside the structure. NFPA 4.7.1. *CO, 2E,* Page 197.

Question #32. As an organization that has a bargaining union, it is important for the company officer to understand the factors that surround the contract and the department so as not to violate any laws or conditions that are legally binding. It is important also to know these as you are the representative for your employees best welfare. NFPA 5.2.1. *CO, 2E,* Page 100.

Question #33. Company officers are expected to lend a hand when needed, whether it is advancing hose or other emergency scene operations so long as it doesn't compromise their role as a leader. The company officer is not setting the mission or developing it. They are the ones working with the crews to meet or accomplish the mission. NFPA 4.2.1. *CO, 2E,* Page 5.

Question #34. Groups exist whenever two or more people share a common goal. Organizations are groups of people. Typically, these people share a common goal, have formal rules, and have designated leaders. This is true of fire companies. NFPA 4.2. *CO, 2E,* Page 44.

Question #35. The recognition of the power of position accompanied by the authority by virtue of the supervisor's ability to administer punishment leads others to see the punishment power concept. NFPA 5.2.1. *CO, 2E,* Page 127.

Question #36. In writing, it is important to remember who is going to read what you wrote. This way, you will write to have effective communications with no barriers. NFPA 4.3.2. *CO, 2E,* Page 28.

Question #37. Leading others is the company officer's primary job as you are the one who is assigned to direct the personnel assigned to you. To accomplish the mission, you must lead others. NFPA 4.2.1. *CO, 2E,* Page 5.

Question #38. Knowing the leading causes of fire as being careless smoking and the evidence presented, you would tend to lean toward that answer based upon fire behavior and the time of day coupled with the ashtray finding and the smoking resident. NFPA 4.5.1. *CO, 2E,* Page 221.

Question #39. Controlling allows us to measure the effectiveness of our effort to help us maintain our goals. By doing so, we can seek ways to improve, thus increasing productivity. NFPA 5.6.1; 5.4.2. *CO, 2E,* Page 79.

Question #40. The first step in the management process is planning. Planning can be for short-, long- or mid-term time frames. NFPA 4.2.2. *CO, 2E,* Page 77.

Question #41. Fire suppression systems like sprinklers were designed to control the fire growth or even suppress the fire. With this assistance, we already have no reported deaths within sprinklered buildings. This system is like giving the fire companies additional personnel for suppression efforts that are housed in each occupancy. NFPA 4.6.2; 4.6.3. *CO, 2E,* Page 254.

Question #42. The first step in the management process is planning and the effectiveness of a company officer is the ability to make and keep plans moving forward. NFPA 5.4.5. *CO, 2E,* Page 77.

Question #43. This is one of several leadership tools that focus the employees toward improving their work performance NFPA 4.2.5. *CO, 2E,* Page 149.

Question #44. The experience of a fire investigator will prove to be beneficial especially with the elements of potential death and potential arson. NFPA 4.5.1; 4.5.2; 5.5.2. *CO, 2E,* Page 277.

Question #45. Most employees described in the question require some discussion, direction, and supervision. They are beginning to contribute ideas and solutions. NFPA 5.2.1, 4.2.1. *CO, 2E,* Page 128.

Question #46. Groups exist whenever two or more people share a common goal. Organizations are groups of people. Typically, these people share a common goal, have formal rules, and have designated leaders. This is true of fire companies. NFPA 4.2. *CO, 2E,* Page 44.

Question #47. The employer should encourage participation and growth along with provide the environment for this to occur. The individual must take advantage of the opportunities to enhance the skills and knowledge for the advancement. NFPA 4.1.1. *CO, 2E,* Page 13.

Question #48. When agreements between employee and employer over some condition are not agreed upon there is a formal process to go through for the dispute. This does protect the employee and employer as they are generally reviewed by a bipartisan group. NFPA 5.2.2. *CO, 2E,* Page 163.

Question #49. Life risk factors are affected by the number of people at risk and their danger and ability to provide for their own safety. There has been no reported fire death in a sprinklered building. NFPA 4.3.1. *CO, 2E,* Page 251.

Question #50. This theory is a management theory introduced by Douglas McGregor. This management style in which the manager believes that people dislike work and cannot be trusted was a traditional style of management. NFPA 5.2.1; 5.2.2. *CO, 2E,* Page 81.

Question #51. You will be a mediator frequently as you try to get others to work out their differences. NFPA 5.7. *CO, 2E,* Page102.

Question #52. It is important to understand the requirements to which you are certifying as a company officer. NFPA 4.1. *CO, 2E,* Page 9.

Question #53. It is important to understand diversity in the areas we work due to the mixture of people and cultures we can encounter. The answer is based upon understanding the population we serve. NFPA 4.3.1. *CO, 2E,* Page 131.

Question #54. The inability to say no or decline the opportunity is the hardest of the time eaters for most. You must remember to tie yourself to goals and focus on priorities. It is important to know when you are overloading your plate. Procrastination is just the opposite. You do not have a focus on goals or priorities as it relates to work. The continued putting off of this project has you in a position of crunch time for completion. This usually leads to poor work. NFPA 5.2.1. *CO, 2E,* Page 106.

Question #55. Whether volunteer or career, the company officer is responsible for the human resources or staffing along with other resources. NFPA 4.2.1; 4.2.2; 4.2.3. *CO, 2E,* Page 54.

Question #56. Hygiene factors are easily controlled by company officers and will enhance the employee's work performance. NFPA 4.7; 5.7 ; 5.2.1. *CO, 2E,* Page 124.

Question #57. All high rise buildings are made of fire-resistive construction. NFPA 4.6.2. *CO, 2E,* Page 248.

Question #58. Principle of scaler organizations is that as you progress with experience and talents, you move upward to higher levels. This is important whereas the company officer is the first level on which individuals accept responsibility for themselves and others. NFPA 4.2. *CO, 2E,* Page 66.

Question #59. A good leader must understand human behavior and the needs of employees as the life outside of the work place often influences life in the work atmosphere. NFPA 4.7; 5.7. *CO, 2E,* Page 123 - 124.

Question #60. As a company officer, it is important to remember to allow in this situation the communications to be a two-way process. NFPA 4.2.2. *CO, 2E,* Page 21.

Question #61. Total quality management principle is a style often practiced by many company officers as they focus on the organizations continuous improvements and keeping customer satisfaction in mind. NFPA 5.4.5. *CO, 2E,* Page 81.

Question #62. It is important to understand the dynamics behind why programs exist within your department and the fire service. As a company officer, you will be faced with questions from the public and youthful firefighters about why the fire service is engaged in prevention and life safety programs. NFPA 4.3.4. *CO, 2E,* Page 217.

Question #63. The inability to say no or decline the opportunity is the hardest of the time eaters for most. You must remember to tie yourself to goals and focus on priorities. It is important to know when you are overloading your plate. NFPA 5.2.1. *CO, 2E,* Page106.

Question #64. It is important to pre-plan situations especially in high-hazard or high-target areas. This will enhance operations and make for safety considerations. NFPA 4.6.1. *CO, 2E,* Page 292.

Question #65. A good leader must understand human behavior and the needs of employees as the life outside of the workplace often influences life in the work atmosphere. NFPA 4.7; 5.7. *CO, 2E,* Page 123.

Question #66. Company officers, for firefighters, are the first step in the chain of command. Command-level officers consider the company officer to be the lower management staff and the first-line supervisors for the organization. NFPA 4.2. *CO, 2E,* Page 4.

Question #67. Knowing the leading causes of fire as being careless smoking and the evidence presented, you would tend to lean toward that answer based upon fire behavior and the time of day, coupled with the ashtray finding and the smoking resident. However, the accidental causes of fires are lead by heating equipment. The question brings forth temperature, time of year, and potentials. NFPA 4.5.1. *CO, 2E,* Page 273.

Question #68. This concept of rising heated gases that fill the space and move downward is a natural process called thermal stratification. NFPA 4.7; 4.6.2. *CO, 2E,* Page 257.

Question #69. Leadership is achieving the organizations goals through others. NFPA 4.2.2. *CO, 2E,* Page 120.

Question #70. The priorities for any incident are as follows in order: life safety, incident stabilization, and property conservation. Operations that make a direct attack on the fire are in the offensive strategy to control the fire and keep it in place. NFPA 4.6.3. *CO, 2E,* Page 327.

Question #71. Life safety is the first priority of emergency operations. This education addresses life safety for the occupants and emergency response procedures. NFPA 5.5.1; 4.3.4. *CO, 2E,* Page 222.

Question #72. This book analyzed the fire problem in America and made recommendations on how to change it. It details many of what we have as common practices today. NFPA 4.4.1; 4.3.4. *CO, 2E,* Page 214.

Question #73. The priorities for any incident are as follows in order: life safety, incident stabilization, and property conservation. NFPA 4.6.2. *CO, 2E,* Page 320.

Question #74. With life safety an incident priority, it is important to understand the pre-planned locations at which these individuals will be located. This will enhance rescue operations. NFPA 4.6.1 & 5.5.2. *CO, 2E,* Page 298.

Question #75. The company officer must be an effective communicator whereas he is giving orders, receiving orders, and dealing both with the public and the crew members. Without effective communications, the company officer can create impressions that are either false or obscured. NFPA 4.2. *CO, 2E,* Page 27.

Question #76. Benchmarking allows command to follow sequential progression in the incident and also note the priorities that are being met. NFPA 4.6.3. *CO, 2E,* Page 320.

Question #77. Action plans address all areas of the fire and in a time sequential order or time frame. NFPA 4.6.3. *CO, 2E,* Page 331.

Question #78. Fire suppression personnel can effectively inspect many occupancies to detect and even facilitate correcting common fire code violations. NFPA 5.5.1; 4.3.4. *CO, 2E,* Page 226.

Question #79. This style of construction has unique fire behavior patterns that will need to be addressed tactically. NFPA 4.6.1. *CO, 2E,* Page 251.

Question #80. The company officer gives many oral commands on emergency scenes and they follow customs, rules, and practices of the industry, thus making them formal. NFPA 4.2.1. *CO, 2E,* Page 20.

Question #81. Because this type of construction allows long spans without support, it is very popular in large one-story commercial facilities. NFPA 4.6.2. *CO, 2E,* Page 249.

Question #82. Supervisors should be sure that all employees are fairly treated and represented in all department activities. NFPA 5.2.1; 4.7.1; 5.2.2. *CO, 2E,* Page 131.

Question #83. Lightweight construction is engineered to be as strong or stronger than solid components. NFPA 4.6.2. *CO, 2E,* Page 250.

Question #84. These are utilized for primary access points barriers to access, utilities, and water supply. They work well for a quick 20-second reference. NFPA 4.6.1. *CO, 2E,* Page 295.

Question #85. Locations where there are unusual hazards or where an incident would overload the department's resources are examples of target hazards. Nursing facilities and retirement centers are high life hazard complexes. NFPA 4.6.1. *CO, 2E,* Page 293.

Question #86. The company officer's job is as varied as all the activities that the organizations does. However, the company officer is the first-line supervisor and generally the senior representative that the public deals with on a routine basis. NFPA 4.1.1. *CO, 2E,* Page 4.

Question #87. The command sequence is a three-step process that helps incident commanders manage the incident. NFPA 4.6.2. *CO, 2E,* Page 327 - 337.

Question #88. Individuals in theory with high achievement needs tend to work more diligently and want to see it as an achievement. NFPA 5.2.2. *CO, 2E,* Page 125.

Question #89. Most organizations have a pyramid structure, with one person in charge, and an increasing number of subordinates at each level as you move downward. The company officer is the first level of supervisor and has a limited number of subordinates for direct supervision, but numbers increase with the size of station and on incidents. NFPA 4.1.1. *CO, 2E,* Page 61.

Question #90. This is when you accept responsibility for others and their actions as a company officer. NFPA 4.2. *CO, 2E,* Page 62.

Question #91. Many cities have developed building and fire codes and went so far as to enforce them before 1871. NFPA 5.5.1. *CO, 2E,* Page 228.

Question #92. Length x Width x Height / 100 = the fire flow for each floor. If the floor is 1/4 involved, then divide that answer by 4 and you would get 1,500 gpm. NFPA 4.6.1. *CO, 2E,* Page 264.

Question #93. Length x Width x Height / 100 = the fire flow for each floor. If the floor is 1/4 involved then divide that answer by 4 and you would get 2,250 gpm. NFPA 4.6.1. *CO, 2E,* Page 264.

Question #94. As the company officer gives orders on the emergency scene, the firefighter, who is the receiver of the message, interprets the message and acts from there. NFPA 4.2.2. *CO, 2E,* Page 21.

Question #95. Safety comes from the elimination of unsafe acts or equipment. It is important as a company officer to recognize the issues and the requirements and standards that support those issues. NFPA 4.7.1. *CO, 2E,* Page 177.

Question #96. The first step in the management process is planning. The effectiveness of a company officer is the ability to make and keep plans moving forward. NFPA 5.4.5. *CO, 2E,* Page 77.

Question #97. In the communications process you will be faced with questions to which you may not know the answer. Be up front and honest that you don't know the answer, but work to find the answer. In doing so, you will enhance your knowledge and develop good communications skills. NFPA 4.2.6. *CO, 2E,* Page 27.

Question #98. This is the document by which you inspect. These codes are adopted by local, state, or other jurisdictional entities. NFPA 5.5.1. *CO, 2E,* Page 228.

Question #99. Attitude is the core of your performance. Your attitude, when positive, will reflect a professional and loyal role model who will be able to lead others to meet the department's mission. NFPA 4.1.1. *CO, 2E,* Page 15.

Question #100. We utilize these barriers as crutches for not delegating work due to our own insecurity in the delegation process. NFPA 4.2.1; 4.2.2. *CO, 2E,* Page 100.

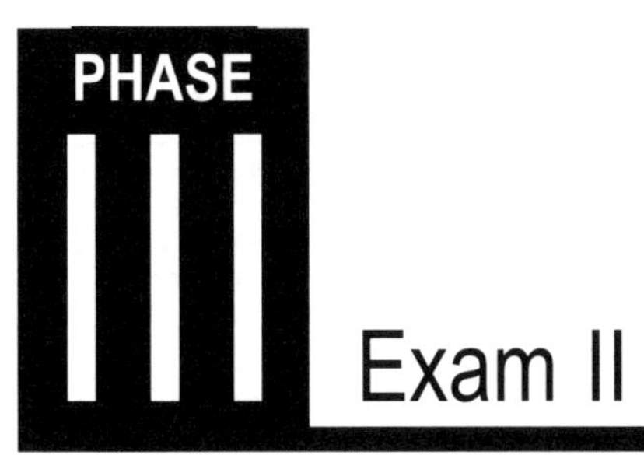

Exam II

1. Pre-incident planning consists of which of the following activities?
   a. pre-incident survey
   b. development of information resources that would be useful during the event
   c. development of procedures that would be used during the event
   d. a & c
   e. all of the above

2. You are about to be formally sworn in as a Lieutenant in your department. Upon this promotion, you are afforded the right and power to command. This right and power by position is known as _______________.
   a. respect
   b. power
   c. managerial legal right
   d. authority

3. The addition of an automatic sprinkler system to a building lowers which of the following?
   a. life risk factors
   b. property risk factors
   c. physical risk factors
   d. community risk factors

4. As a company officer, in which of the following three essential elements in fire prevention would you be the most involved?
   a. education
   b. engineering
   c. enforcement

5. You are a newly promoted lieutenant and you have been assigned personnel. These personnel are dependant upon you for guidance daily. This makes your primary role _______________.
   a. leadership
   b. coaching
   c. mentoring
   d. friendship

6. As a company officer, in which of the following three essential elements in fire prevention would you be the least involved ______________.
   a. education
   b. engineering
   c. enforcement

7. As a company officer it is necessary to provide a good work environment for your employees. Herzberg's model describes what is needed to get employees to come to work and not be dissatisfied. Items like good working relationships and considerate supervisors are examples of ______________.
   a. hygiene factors
   b. motivators
   c. achievements
   d. power

8. Use the Iowa State formula. You respond to a commercial structure which is 200 feet wide, 400 feet long, and 10 feet high on each of 20 floors. With this knowledge, figure fire flow based on a appropriate theoretical fire flow formula utilizing all of the building's dimensions. How much would the fire flow be (in gpms) if the building was 100% involved on the 8th floor and 25% involved on the 9th floor?
   a. 10, 000 gpm
   b. 1,500 gpm
   c. 3,000 gpm
   d. 30,000 gpm

9. You have left the scene of an apartment fire. You are presented with an anonymous tip of information not found on the initial investigation. Which of the following are ways the fire department may reenter the property after they have left the scene?
   a. administrative search warrant
   b. administrative process
   c. owner's consent
   d. search warrant
   e. all of the above

10. As a company officer, you will receive a larger work load and responsibility with the position. Often times, tasks are assigned to the company officer to complete. As work loads increase, it is often easy to become overloaded. This is due to the officer not delegating appropriate assignments and trying to do the entire assignment themselves. This occurs when officers will say "I can do it better myself," "It will take to much time to train someone to do that task," or I really can't take a chance on that project." These are examples of ______________.
    a. barriers to delegation
    b. reasons for delegation
    c. hemisphere of delegation
    d. facts about delegation

11. Company officers have status and power. We add this concept to our list of motivators that get employees to do what is asked of them. You ask a firefighter to wax the kitchen floor. The task is quickly completed. One reason would be that you, as the company officer, have ______ power.
    a. legitimate
    b. reward
    c. punishment
    d. identification

12. You have been a company officer for 10 years. During your first couple of years, you struggled with being a meets expectation employee. Over the last 8 years, you have been found to be reliable and dependable to the higher administrative staff on assigned projects. The more you have proven your talents, the more special assignments you seem to be getting pushed your way. Now you are finding that you are struggling to get routine tasks accomplished again. Which one of the following best describes you in the proven time eaters?
    a. lack of personal goals
    b. reacting to urgent events
    c. inability to say no
    d. trying to do too much yourself

13. You are the company officer and one of your employees feels that your disciplinary actions are too harsh for him not meeting a deadline on a project. You and the employee are unable to come to an agreement even with formal meetings. You would expect the employee to start _______________.
    a. gripe
    b. grievance
    c. conflict
    d. complaint

14. A company officer who is conducting an in-service fire inspection can enforce which of the following?
    a. fire prevention code
    b. electrical code
    c. building Code
    d. a & c
    e. a & b

15. You have a large retirement facility with skilled nursing facility included and an assembly that has been built recently in your jurisdiction. These types of complexes are examples of _______________.
    a. fire targets
    b. target hazards
    c. facilities with multiple automatic alarms daily
    d. none of the above

16. Place in order the following pieces of the command sequence: 1. size-up 2. implementing an action plan 3. developing an action plan
    a. 1, 3, 2
    b. 3, 2, 1
    c. 2, 3, 1
    d. 1,2, 3

17. The span of control is important as a company officer is responsible for human resources. According to the Company Officer textbook, the number of personnel that a supervisor can optimally manage is _____ to ______.
    a. 2, 4
    b. 4, 7
    c. 3, 5
    d. 5, 8

18. You are assigned as the lieutenant of a task force that will be working in a collapsed building due to an explosion of natural gas. You have selected team members that are specifically trained in the area of structural collapse with specialties in shoring, engineering design, and fiber optic visual technology. As the task force leader, you would be considered to be___________ this group as you begin rescue operations for trapped victims.
    a. organizing
    b. commanding
    c. developing
    d. coordinating

19. As a company officer, effective communications will _______________. 1. enhance leadership ability 2. gain respect from peers and superiors 3. assist in dealing with the media
    a. 1 & 2
    b. 2 & 3
    c. 1, 2 & 3
    d. 1 only
    e. 3 only

20. As a company officer, you have assigned members of your crew to do preplanning activities. This is a new program the department is trying to get established. You have acquired outside training in the subject matter from an accredited institution. You have spent considerable time training the crew members and have given instructions and made assignments. What is described here is an example of _______________.
    a. organization
    b. coordination
    c. management
    d. leadership

21. You are the company officer and the first arriving company officer to the scene of a multi-family residential apartment complex on Good Friday morning. You arrive to find this two story apartment complex with heavy fire involvement on the front of the building with three floors involved and at least four apartments. Upon exiting your apparatus, you begin asking occupants if everyone is out of the building. They feel sure some are missing. As you come around the front of the apparatus, you hear a man's voice screaming for help, stating he is on the third floor. Then the voice stops. Which of the following sequences best describes your action priorities?
    a. life safety, incident stabilization, property conservation
    b. incident stabilization, property conservation, life safety
    c. life safety, property conservation, incident stabilization
    d. a or b
    e. all of the above

22. You are the company officer determining an employee's competency and commitment. This employee has weak skills and good self-confidence. They are always looking for ways to get out of work or avoid projects. This employee would be which of the following?
    a. high competency / high commitment
    b. low competency / high commitment
    c. high competency / low commitment
    d. low competency / low commitment

23. As the company officer, you will use many leadership styles to accomplish the mission of the department and the goals of the company. You have an employee who is a five year veteran and has demonstrated strength in skills and abilities on the job. This employee functions at the next level consistently and routinely takes on projects and requires little support from you. Which leadership style will you most likely utilize?
    a. directing
    b. consulting
    c. supporting
    d. delegating

24. You are the company officer and an employee comes to you with a problem. They state they have a problem, but also have a potential solution. To assure you have received the complete message, you should practice which of the following?
    a. communications
    b. barrier breakdown
    c. active listening
    d. passive listening

25. You are a company officer who is in charge of a crew who is stationed in a mixed district representing some business and residential response potential. In the response district, the residential configurations range from million dollar single family homes to multi-family housing to mobile homes. This mixture will bring a diverse group of residents that depend on you for services. To understand all of the population, it is important to have a crew strong in which of the following areas?
    a. emergency scene operations
    b. cultural diversity
    c. tactical considerations
    d. all of the above

26. Apparatus and activities are ongoing on the emergency scene. These pieces are considered to be which of the following in the communications process?
    a. message
    b. barriers
    c. feedback
    d. transmitter

27. For what does the "W" in "COAL WAS WEALTH" stand?
    a. water
    b. weather
    c. waste
    d. warranted

28. If the fire chief would come by a station to have a friendly lunch with the station personnel this would be an example of a(n) _______________.
    a. informal group
    b. formal group
    c. formal relationship
    d. informal relationship

29. You respond to the Old Main Street portion of the city for a reported fire in a five-story commercial building. This portion of the city was mostly constructed in the 1920s. Which type of building construction would you anticipate to see?
    a. fire-resistive
    b. non-combustible
    c. heavy timber
    d. ordinary

30. You are on the third alarm responding to a large industrial building fire. Knowing you will be the first engine arriving on the scene, in the second alarm, the incident commander summons you to the command post and assigns your crew to work as part of a task force with another company. You are given the task of tracking personnel and their location during the incident. This process is known as ______________.

    a. incident command
    b. unified command
    c. personnel accountability
    d. personnel management

31. You have an employee who brings in a pornographic picture to work. He shows a picture of a female in the magazine to several members of the crew. He states the female in the picture looks like another department employee. What answer best describes the condition described here?

    a. harassment
    b. sexual harassment
    c. Civil Rights Act violation
    d. b & c
    e. all of the above

32. The definition of ________ is being responsible for one's personal activities; in the organizational context, this includes being responsible for the actions of one's subordinates.

    a. responsibility
    b. accountability
    c. line authority
    d. management

33. As a company officer, you are always looking for the advantage over emergency responses especially fire. Which one of the following would be classified as one of our greatest allies in fire protection?

    a. construction features
    b. fire codes
    c. sprinkler systems
    d. smoke detectors

34. You as the company officer have been instructed by the fire chief to write a letter to a citizen referencing a complaint about the use of sirens during the middle of the night. Which of the following is important?

    a. technical terms
    b. emphasis
    c. considering the reader
    d. complexity

35. As a company officer, you must look at the needs for your crew and develop them accordingly. The first step in this management process would be ______________.
    a. planning
    b. delegating
    c. centralizing
    d. advocating

36. You have an employee who you feel doesn't like work and you have found that he cannot be trusted to keep deadlines or meet minimum productivity levels. You are describing which of the following management theories?
    a. TQM
    b. Theory X
    c. Theory Y
    d. Theory Z

37. As a company officer, you have the ability to get many tasks accomplished. When promoted, you were automatically given power through the position. By this power, you make employees work to levels to which they may not normally be self-motivated to work. This concept is known as ______________ power.
    a. reward
    b. legitimate
    c. punishment
    d. persuasion

38. As a company officer, you are faced with a question for which you are not sure of the answer. What should you do?
    a. Answer that you are busy and research would help them learn.
    b. Admit you don't know the answer.
    c. Try to find the answer.
    d. a & b
    e. b & c

39. You arrive on the fire scene and give a size-up. Upon entering the structure, you are advising command you are entering with six personnel. When you begin your attack, you announce water on the fire. These are examples of ______________.
    a. tactics
    b. strategies
    c. flow procedures
    d. benchmarks

40. Your preplans program is underway and assignments have been made. You are instructed to use floor-by-floor drawing on the pre-incident survey sheet. Which of the following best describes the drawing portion?
    a. bird's-eye view of the property
    b. floor plan
    c. engineering schematic
    d. three-dimensional plan

41. You are inside a working structure fire and you encounter heat that is significant in nature near the three-foot mark off of the floor. Below that mark, it is significantly cooler. In fire behavior, you know that the heat is an indication of the potentials that exist. We know this concept on higher temperatures closer to the ceiling as _______________.
    a. thermal stratification
    b. line of demarcation
    c. thermal plane
    d. thermal balance

42. You respond to a residential structure fire that has been vacant over the last six months. The electrical meter base is capped with no meter in place. The fire started on the second floor. The rear door is open and you find strange patterns of char on the floor with close areas not burned. What is your assumption on the cause of this fire?
    a. electrical cord under the rugs
    b. arson
    c. electrical wiring fire in the floor
    d. none of the above

43. A crew in a fire department is which of the following?
    a. formal organization
    b. informal organization
    c. formal organization within an informal organization
    d. all of the above

44. Acme fire department has a progressive career development path for individuals to move up through the ranks from firefighter to chief. This uninterrupted series of steps or layers is an organizational concept known as _______________.
    a. linear principle
    b. scaler principle
    c. unity of command
    d. division of labor

45. The publication that was published in 1987 that included 90 recommendations that could help reduce fire loss in America was________.
    a. American Firefighter
    b. America Burned
    c. America Burning
    d. America Burning Revisited

46. You are senior company officer who has been tasked to look into the health and safety program of the department. You are tasked with bringing the department's rules, regulations, and standard operating guidelines into compliance with professional standards and laws. Which two organizations would you look to for information and guidance?
    a. JEMS and FIREHOUSE Magazines
    b. NIOSH and JOCCA
    c. NFPA and National Fire Rescue Magazine
    d. NFPA and OSHA

47. Old woodframe structures which were built in the early 1900s and late 1800s will probably be built in a style known as _______________.
    a. platform
    b. old school
    c. oak frame
    d. balloon frame

48. The term company is used to describe work teams. The individual of that work team who must coordinate activities is the company officer. This individual has many roles to fill. Which of the following is not one of those roles?
    a. help in advancing hose
    b. directing crew operations
    c. activities that don't compromise roles as leaders
    d. developing a mission statement

49. If you are preplanning an old cotton mill, you would anticipate the construction to be _______________.
    a. ordinary
    b. Type III
    c. heavy timber
    d. Type V
    e. none of the above

50. You have an employee who is working hard and not making great progression in their career. Everything in their life seems to be falling apart to include a divorce and loss of a house. This employee is just at work going through the motions most of the time. According to Maslow, this individual would be in which position in his hierarchy of needs model?
    a. social needs
    b. security needs
    c. self-actualization needs
    d. survival mode

51. During emergency scene operations, the lieutenant typically would be the _______ in the communications model.
    a. sender
    b. message
    c. receiver
    d. all of the above

52. Under the needs theory, a firefighter who you supervise takes responsibility for their efforts and has far exceeded the training requirements for two positions above their position. They have set even higher goals with you to continue to grow. You would say this person is which of the following?
    a. high-achievement needs
    b. high-affiliation needs
    c. low-achievement needs
    d. low-affiliation needs

53. You are the senior company officer and the first arriving company officer to the scene of a multi-family residential apartment complex on Christmas night. You arrive to find this three-story apartment complex with heavy fire involvement on the first and second floors of one end involving at least four apartments. Dispatch had a subject on the phone begging for help for his family when the phone was dropped and the line remained open for a period of time before going dead. You have several victims unaccounted for at this time. You are making an interior attack. This would be known as _______________ mode.
    a. offensive
    b. defensive
    c. transitional
    d. none of the above

54. As a captain of a company, which of the following three essential elements in fire prevention would you be the most involved in if you were doing in-service company inspections in a strip mall?
    a. education
    b. engineering
    c. enforcement

55. The company officer is the crew leader and first-line supervisor of assigned staffing and equipment. We refer to these assigned field personnel as _______________.
    a. staff
    b. bodies
    c. human resources
    d. facility resources

56. As a company officer, you will set levels of achievement with your employees that can be effectively measured to assure that achievement has occurred. In setting these levels, you are helping the employee set obtainable ______________.
    a. objectives
    b. JPRs
    c. progress
    d. goals

57. As a company officer, you will open channels of communications between employees and other department ranks with your crew members. In doing so, you often persuade the two sides to meet to discuss and work out differences. This concept is known as ______________.
    a. meditation
    b. arbitration
    c. mediation
    d. addiction

58. As the company officer, you will use many leadership styles to accomplish the mission of the department and the goals of the company. You have an employee who is fresh out of the academy. Which leadership style will you most likely utilize?
    a. directing
    b. consulting
    c. supporting
    d. delegating

59. You are a company officer who arrives on the scene of a mall complex. You see heavy volumes of fire coming through the roof of an anchor store. as which type of building construction would this strip mall be classified?
    a. fire-resistive
    b. non-combustible
    c. heavy timber
    d. frame

60. Orders given on the emergency scene during operations by the company officer are in the form of ______________.
    a. informal
    b. formal
    c. oral
    d. a & c
    e. b & c

61. You are the senior company officer and you have been given a house in which to do live fire training. You will be following NFPA 1403. Your structure is a single-story house with 2,210 square feet. It has a large fireplace in the living room and a basement with narrow steps leading down off of the kitchen. What will be required to make this structure safe?
    a. floors, railing, and stairs made safe
    b. asbestos removed
    c. chimney hazards eliminated
    d. extraordinary dead weight removed
    e. all of the above

62. Career development is a shared responsibility between which of the following?
    a. the individual and the community
    b. the individual and the department
    c. the community and the department
    d. the environment and you

63. As the company officer, you will use many leadership styles to accomplish the mission of the department and the goals of the company. You have an employee who is a five-year veteran and has demonstrated progressing strength in skills and abilities on the job. Which leadership style will you most likely utilize?
    a. directing
    b. consulting
    c. supporting
    d. delegating

64. You arrive on scene of a two-story woodframe single residential structure that you were dispatched to for an automatic fire alarm. Upon entering the structure you find a light haze of smoke on the first floor. Further investigation leads you to a smoldering waste basket in the bathroom. This fire was in which of the following stages?
    a. incipient
    b. free-burning
    c. backdraft
    d. smoldering

65. You are reviewing an employee's performance evaluation for the quarter. It is important to have good communications occurring. This two-way communications allows the employee to provide _______________ on the evaluation.
    a. feedback
    b. knowledge
    c. barriers
    d. all of the above

66. You are the investigating officer on a small room and contents fire in a bedroom. The fire began about 11PM on Friday, August 21. You have conducted interviews with the occupants who are smokers and are now looking at findings in the room. The bed is burned significantly and the heat line of demarcation is about 5 feet off the floor. There is heavy smoke residue in the room with minor fire damage to any other furniture.You find a space heater plugged up close to the window beside the bed, several extension cords, and an ashtray by the bed. What would you suspect the cause of this fire to be?
    a. electrical
    b. improperly positioned space heater
    c. smoking
    d. not enough information to determine

67. You respond to an automatic alarm and are the second apparatus to arrive. Your standard assignment is to take the water supply and the fire department connection. You do this based on the procedures that are written known as which of the following?
    a. quick access pre-plans
    b. notes
    c. standing operating procedures
    d. none of the above

68. As a newly promoted company officer, you are _______________.
    a. responsible for the performance of the assigned crew
    b. responsible for the safety of the assigned crew
    c. the second-line supervisor
    d. a & b
    e. all of the above

69. A company officer certified as a Fire Officer I should meet the National Fire Protection Association Standard 1021 ________.
    a. Chapter 2
    b. Chapter 3
    c. Chapter 4
    d. Chapter 5

70. You are the investigating officer on a small room and contents fire in a bedroom. The fire began about 11 PM on Friday, February 1. The temperature outside is -10 degrees and there is 8 inches of snow on the ground. You have conducted interviews with the occupants who are smokers and are now looking at findings in the room. The bed is burned significantly and the heat line of demarcation is about 5 feet off the floor. There is heavy smoke residue in the room with minor fire damage to any other furniture.You find a space heater plugged in close to the window beside the bed, several extension cords, and an ashtray by the bed. What would you suspect the cause of this fire to be?
    a. electrical
    b. improperly positioned space heater
    c. smoking
    d. not enough information to determine

71. The ABC fire department has a pyramid organizational structure. Where does the company officer fall within this structure?

    a. front-line supervisor
    b. first-line supervisor
    c. second-level supervisor
    d. a & b
    e. a & c

72. As an applicant with a department, your employment is conditional on your joining the union. Failure to join or agreement to join resulting in your employment is an example of a(n) ______________ shop.

    a. open
    b. closed
    c. union
    d. agency

73. You are the senior company officer and the first arriving company officer to the scene of a multi-family residential apartment complex on Halloween night. You arrive to find this two-story apartment complex with heavy fire involvement on the first and second floors of one end involving at least four apartments. Dispatch had a subject on the phone begging for help when the phone was dropped and the line remained open for a period of time before going dead. You now have brought the fire under control. You have a victim unaccounted for at this time Which of the following meet the guidelines for calling a fire Investigator?

    a. potentially incendiary fire
    b. a fire resulting in a death
    c. property damage greater than $50,000
    d. a & b
    e. all of the above

74. You are a company officer in a department that has multiple stations. Your station is a multiple-company station and you are a seasoned company officer on a rescue company. Your battalion chief asks you to develop a five-year station work plan. What type of plan would you be developing?

    a. long-range term
    b. mid-range plan
    c. short- range plan
    d. mini plan

75. As a company officer, you meet with your crew and set a vision for where the company is going based on the department's mission statement. You set clear concise expectations based upon the guiding principles of the department. In developing this, you are focusing on both the internal and external customer's needs. In doing this, you are practicing what management style?

    a. Theory X
    b. Theory Z
    c. traditional management
    d. total quality management

76. You are the company officer determining an employee's competency and commitment. This employee has strong skills and self-confidence. They are always looking for ways to get involved or projects on which to work. This employee would be which of the following?

    a. high competency / high commitment
    b. low competency / high commitment
    c. high competency / low commitment
    d. low competency / low commitment

77. You are the company officer determining an employee's competency and commitment. This employee has strong skills and self-confidence. They are always looking for ways to get out of work or avoid projects. This employee would be which of the following?

    a. high competency / high commitment
    b. low competency / high commitment
    c. high competency / low commitment
    d. low competency / low commitment

78. A volunteer fire department would be which of the following?

    a. formal organization
    b. informal organization
    c. combination organization
    d. all of the above

79. Lightweight scissor wood trusses are commonly used in which of the following types of construction?

    a. Type I
    b. Type II
    c. Type III
    d. Type IV
    e. Type V

80. As a company officer, you are issued a legal document that sets forth the requirements for life safety and property protection in the event of fire, explosion, or similar emergency. Its purpose is to minimize the risk of loss of life and property by regulating the use and storage of materials that might be on the property. This legal document is which of the following?
    a. fire prevention code
    b. building code
    c. minimal standards
    d. elements and standards

81. You have been a company officer for five years and have been found to be reliable and dependable to the higher administrative staff on smaller assigned projects. The chief has asked you to head up a hiring process for new recruits. It is summer time and you have a new boat you and your family are enjoying everyday you are off. You also have taken several days off to go to the lake. You were assigned the hiring project on June 9. The completion date has been set for July 23. Today it is July 13 and you have done very little work on this project. Which one of the following best describes you in the proven time eaters?
    a. lack of personal goals
    b. reacting to urgent events
    c. procrastination
    d. trying to do too much yourself

82. Professionalism encompasses attitude, behavior, communication, style, demeanor, and ethical beliefs. Which of the following best describes attitude?
    a. the core of your performance
    b. nothing that helps job performance
    c. suggests you are working to be a good role model
    d. all of the above
    e. a & c

83. You are a company officer in a department that has multiple stations. Your station is a multiple-company station and you are a newly promoted company officer on a rescue company. Your battalion chief asks you to develop a work plan for the next year. What type of plan would you be developing?
    a. long-range term
    b. mid-range plan
    c. short- range plan
    d. mini plan

84. You are the company officer of an employee who has thrown a fire cracker in the day room where there are several other employees. The fire cracker detonated. This caused one of the employees to have some pain in their ear. You are going to do take disciplinary action on the employee's behavior. Which of the following would be the most appropriate disciplinary form?
   a. transfer
   b. demotion
   c. oral reprimand
   d. termination

85. You are the company officer of an employee who has thrown a fire cracker in the day room where there are several other employees. The fire cracker detonated. This caused one of the employees to have some pain in their ear. Which of the following would be the most appropriate answer?
   a. applauding the employee
   b. coaching the employee
   c. counseling the employee
   d. ignoring the situation; it was a joke

86. You respond to a commercial structure which is 100 feet wide, 400 feet long, and 10 feet high. With this knowledge, you could figure fire flow based on which theoretical fire flow formula?
   a. National Fire Academy Formula
   b. Iowa State University Formula
   c. West Virginia University Formula
   d. a & b
   e. a & c

87. Your company has been assigned to a city festival in the fire department's booth with life safety educators. Your interaction and primary responsibilities for the assignment would be _______________.
   a. tactical
   b. standing by
   c. fire suppression
   d. fire prevention

88. Use the Iowa State formula. You respond to a commercial structure which is 300 feet wide, 500 feet long, and 12 feet high on each of 50 floors. With this knowledge, figure fire flow based on an appropriate theoretical fire flow formula utilizing all of the building's dimensions. How much would the fire flow be (in gpms) if the building was 25% involved on the 50th floor?
   a. 450,000 gpm
   b. 450 gpm
   c. 4,500 gpm
   d. 45,000 gpm

89. You are the company officer who arrives on scene of a 100-story building with fire showing from one window on the 22nd floor. The building described here will most likely be made of which type of construction?

    a. fire-resistive

    b. non-combustible

    c. woodframe

    d. ordinary

90. You are the senior company officer and find yourself in the command role. You have companies that are attacking the fire, addressing exposure, ventilation, water supply, and rescue. With this, you have organized a course of action that addresses all aspects for controlling the incident. This is known as a _______________.

    a. plan of accord

    b. action plan

    c. resource plan

    d. action series

91. You are a newly promoted lieutenant. From where will most of your management occur?

    a. behind a desk

    b. in the field

    c. in the fire station

    d. none of the above

92. As a company officer, you are delegated a project to develop a public education prop for fire prevention week. You are to make a prop that resembles a fire truck for a puppet show. You utilize one of your crew member's talent in carpentry. This concept is known as _______________.

    a. delegation

    b. advocation

    c. management

    d. supervision

93. Which one of the following has the greatest monetary loss attributed to its damages annually?

    a. hurricanes

    b. fire

    c. tornadoes

    d. floods

94. You have an employee who has just gone through the death of their spouse leaving them with two children and a large sum of bills due to a long illness. This employee has spent countless hours caring for their spouse and continuing to work and care for the children. Which of the following best describes the position in which this person would fall in Maslow's hierarchy of needs?
    a. self-actualization
    b. self-esteem
    c. security needs
    d. social needs

95. You are a company officer in a department that has multiple stations. Your station is a multiple-company station and you are a seasoned company officer on a rescue company. Your battalion chief asks you to develop a three-year station work plan. One year into this plan, you find that you are behind by two months. During the next year, you make up this deficit and are on schedule as planned. This is an example of _______________.
    a. controlling
    b. commanding
    c. organizing
    d. developing

96. As a company officer, you may be assigned to manage the department's health and safety program. If you had this assignment, you would have the responsibility and title of _______________.
    a. safety officer
    b. health and safety
    c. health and safety officer
    d. none of the above

97. You and your company are the dispatched unit that command has designated as the rapid intervention team. Which of the following best describes your duties per the textbook?
    a. You are the team of firefighters which will have a minimum of two personnel. You are assigned the primary task of rescuing firefighters should the need arise. You will also be immediately available for any task that command has to assign to you.
    b. You are the team of firefighters which will have a minimum of two personnel. You are assigned the primary task of rescuing firefighters should the need arise. You will not be immediately available for any task that command has to assign to you.

98. Certification means that an individual has been tested by an accredited examining body on clearly identified materials and found to meet the minimal standard. There are other reasons to certify. Which of the following is not one of the reasons to certify?
    a. protection from liability
    b. recognition of demonstrated proficiency
    c. recognition of professionalism
    d. all of the above
    e. b & c only

99. You are a newly promoted lieutenant in your department. Where do you fit into the department's organization?
    a. middle management
    b. lower management
    c. first-line supervisor
    d. a & c
    e. b & c

100. You respond to a reported structure fire in a university dormitory where there are known invalids. The plan for them is to protect in place or go to a portion of the structure that is relatively safe from fire and the by-products of fire. This area is known as an area of _______________.
    a. security
    b. safety
    c. congregating
    d. refuge

Phase III, Exam II: Answers to Questions

1. E
2. D
3. A
4. A
5. A
6. B
7. A
8. A
9. E
10. A
11. C
12. C
13. B
14. D
15. B
16. A
17. B
18. B
19. C
20. D
21. A
22. B
23. D
24. C
25. B
26. B
27. B
28. D
29. D
30. C
31. E
32. B
33. C
34. C
35. A
36. B
37. B
38. E
39. D
40. B
41. A
42. B
43. A
44. B
45. D
46. D
47. D
48. D
49. C
50. D
51. A
52. A
53. A
54. C
55. C
56. D
57. C
58. A
59. B
60. E
61. E
62. B
63. B
64. A
65. A
66. C
67. C
68. D
69. C
70. B
71. D
72. B
73. E
74. B
75. D
76. A
77. C
78. A
79. E
80. A
81. C
82. A
83. C
84. C
85. C
86. D
87. D
88. C
89. A
90. B
91. B
92. C
93. B
94. C
95. A
96. C
97. B
98. D
99. E
100. D

Phase III, Exam II: Rationale & References for Questions

Question #1. It is important to pre-plan situations especially in high hazard or high target areas. This will enhance operations and safety considerations. NFPA 4.6.1. *CO, 2E,* Page 292.

Question #2. The position of rank affords the right and the power to command. This is the authority you are granted as a company officer. NFPA 4.4.1. *CO, 2E,* Page 99.

Question #3. Life risk factors are affected by the number of people at risk and their danger and ability to provide for their own safety. There has been no reported fire death in a sprinklered building. NFPA 4.3.1. *CO, 2E,* Page 253.

Question #4. Life safety is the first priority of emergency operations. This education addresses life safety for the occupants and emergency response procedures. NFPA 5.5.1; 4.3.4. *CO, 2E,* Page 222.

Question #5. Leading others is the company officer's primary job as you are the one who is assigned to direct the personnel assigned to you. To accomplish the mission, you must lead others. NFPA 4.2.1. *CO, 2E,* Page 5.

Question #6. This area is generally left to fire protection engineers and building code officials to assure they meet standards and will protect the structure. NFPA 5.5.1; 4.3.4. *CO, 2E,* Page 223.

Question #7. Hygiene factors are easily controlled by company officers and will enhance the employee's work performance. NFPA 4.7; 5.7 ; 5.2.1. *CO, 2E,* Page 124.

Question #8. Length x Width x Height / 100 = the fire flow for each floor. If the floor is 1/4 involved, then divide that answer by 4. Then, add both numbers together to get the amount of water needed. NFPA 4.6.1. *CO, 2E,* Page 264.

Question #9. Knowing the legal system and the required process is a must if this situation was to occur. NFPA 4.5.1; 4.5.2; 5.5.2. *CO, 2E,* Page 281.

Question #10. We utilize these barriers as crutches for not delegating work due to our own insecurity in the delegation process. NFPA 4.2.1; 4.2.2. *CO, 2E,* Page 100.

Question #11. The recognition of the power of position accompanied by the authority by virtue of the supervisor's ability to administer punishment leads others to see the punishment power concept. NFPA 5.2.1. *CO, 2E,* Page 127.

Question #12. The inability to say no or decline the opportunity is the hardest of the time eaters for most. You must remember to tie yourself to goals and focus on priorities. It is important to know when you are overloading your plate. NFPA 5.2.1. *CO, 2E,* Page 106.

Question #13. When agreements between employee and employer over some condition are not agreed upon there is a formal process to go through for the dispute. This does protect the employee and employer as they are generally reviewed by a bipartisan group. NFPA 5.2.2. *CO, 2E,* Page 163.

Question #14. Many cities have developed building and fire codes and went so far as to enforce them before 1871. NFPA 5.5.1. *CO, 2E,* Page 228.

Question #15. Locations where there are unusual hazards or where an incident would overload the department's resources are examples of target hazards. Nursing facilities and retirement centers are high life hazard complexes. NFPA 4.6.1. *CO, 2E,* Page 293.

Question #16. The command sequence is a three-step process that helps incident commanders manage the incident. NFPA 4.6.2. *CO, 2E,* Page 327 - 337.

Question #17. As a company officer, it is important to be able to manage incidents and daily functions. The use of span of control is important to be able to effectively manage these personnel. NFPA 4.2. *CO, 2E,* Page 67.

Question #18. The third step in the management process, commanding, involves using the talents of others, giving them directions, and setting them to work. NFPA 5.6.1. *CO, 2E,* Page 78.

Question #19. The company officer must be an effective communicator whereas he is giving orders, receiving orders, and dealing both with the public and the crew members. Without effective communications, the company officer can create impressions that are either false or obscured. NFPA 4.2. *CO, 2E,* Page 27.

Question #20. Leadership is achieving the organization's goals through others. NFPA 4.2.2. *CO, 2E,* Page 120.

Question #21. The priorities for any incident are as follows in order: life safety, incident stabilization, and property conservation. NFPA 4.6.2. *CO, 2E,* Page 320.

Question #22. Understanding an employee's competency and commitment is important as you mentor and work with these individuals as a supervisor. NFPA 4.2.2. *CO, 2E,* Page 146.

Question #23. Most employees described in the question require little discussion, direction, and supervision. They are at the next level waiting for challenges and opportunities to present themselves. NFPA 5.2.1, 4.2.1. *CO, 2E,* Page 128.

Question #24. The deliberate and apparent process by which one focuses attention on the communications of another is active listening. This is important to the company officer so as to get the full message of the individual presenting the problem and solution. NFPA 4.2.2. *CO, 2E,* Page 24.

Question #25. It is important to understand diversity in the areas we work due to the mixture of people and cultures we can encounter. The answer is based upon understanding the population we serve. NFPA 4.3.1. *CO, 2E,* Page 131.

Question #26. There are many barriers or obstacles to overcome on emergency scenes. Noise from these are the ones that exist on the most frequent basis. NFPA 4.2.1, 4.2.2. *CO, 2E,* Page 23.

Question #27. This acronym helps with the size-up process. NFPA 4.6.2. *CO, 2E,* Page 326.

Question #28. Although the fire service is a formal organization, informal relationships exist as well, just as described in this question where a superior officer has an informal lunch with the crew of a station. NFPA 4.2.1; 4.2.2; 4.2.3. *CO, 2E,* Page 60.

Question #29. This style of construction is referred to as Main Street, USA. NFPA 4.6.2. *CO, 2E,* Page 249.

Question #30. NFPA 1500 provides policies for managing events to include accountability. NFPA 4.7.1; 4.6.3; 5.6.1. *CO, 2E,* Page 197.

Question #31. Supervisors should be sure that all employees are fairly treated and represented in all department activities. NFPA 5.2.1; 4.7.1; 5.2.2. *CO, 2E,* Page 131.

Question #32. This is when, as a company officer, you accept responsibility for others and their actions. NFPA 4.2. *CO, 2E,* Page 62.

Question #33. Fire suppression systems like sprinklers were designed to control the fire growth or even suppress the fire. With this assistance, we already have no reported deaths within sprinklered buildings. This system is like giving the fire companies additional personnel for suppression efforts that are housed in each occupancy. NFPA 4.6.2; 4.6.3. *CO, 2E,* Page 254.

Question #34. In writing, it is important to remember who is going to read what you wrote. This way, you will write to have effective communications with no barriers. NFPA 4.3.2. *CO, 2E,* Page 28.

Question #35. The first step in the management process is planning. Planning can be for short-, long-, or mid-term time frames. NFPA 4.2.2. *CO, 2E,* Page 77.

Question #36. This theory is a management theory introduced by Douglas McGregor. This management style in which the manager believes that people dislike work and cannot be trusted was a traditional style of management. NFPA 5.2.1; 5.2.2. *CO, 2E,* Page 81.

Question #37. Company officers have a status and power. We add this concept to our list of motivators that get employees to do what is asked of them. NFPA 5.2.1. *CO, 2E,* Page 127.

Question #38. In the communications process you will be faced with questions to which you may not know the answer. Be up front and honest that you don't know the answer, but work to find the answer. In doing so, you will enhance your knowledge and develop good communications skills. NFPA 4.2.6. *CO, 2E,* Page 27.

Question #39. Benchmarking allows command to follow sequential progression in the incident and also note the priorities that are being met. NFPA 4.6.3. *CO, 2E,* Page 320.

Question #40. These are utilized to show major internal structural features. NFPA 4.6.1. *CO, 2E,* Page 295.

Question #41. This concept of rising heated gases that fill the space and move downward is a natural process called thermal stratification. NFPA 4.7; 4.6.2. *CO, 2E,* Page 257.

Question #42. In the situation described, with an unoccupied structure without power, there is little chance of an electrical fire. Arson is motivated by spite, fraud, intimidation, and concealment of a crime, etc. NFPA 4.5.1. *CO, 2E,* Page 274.

Question #43. Groups exist whenever two or more people share a common goal. Organizations are groups of people. Typically, these people share a common goal, have formal rules, and have designated leaders. This is true of fire companies. NFPA 4.2. *CO, 2E,* Page 44.

Question #44. Principle of scaler organizations is that as you progress with experience and talents, you move upward to higher levels. This is important whereas the company officer is the first level at which individuals accept responsibility for themselves and others. NFPA 4.2. *CO, 2E,* Page 66.

Question #45. This book initially analyzed the fire problem in America and made recommendations on how to change it. It illustrates many of what we have as common practices today. America Burning Revisited was to measure the progress being made in the last book's findings. NFPA 4.4.1; 4.3.4. *CO, 2E,* Page 214.

Question #46. Safety comes from the elimination of unsafe acts or equipment. It is important as a company officer to recognize the issues and the requirements and standards that support those issues. NFPA 4.7.1. *CO, 2E,* Page 177.

Question #47. This style of construction has unique fire behavior patterns that will need to be addressed tactically. NFPA 4.6.1. *CO, 2E,* Page 251.

Question #48. Company officers are expected to lend a hand when needed, whether it is advancing hose or other emergency scene operations so long as it doesn't compromise role as a leader. The company officer is not setting the mission or developing it. They are the ones working with the crews to meet or accomplish it. NFPA 4.2.1. *CO, 2E,* Page 5.

Question #49. These structure are called Mill structures or heavy timber based on the fact that the majority of these structures that were built were mills. NFPA 4.6.2. *CO, 2E,* Page 250.

Question #50. A good leader must understand human behavior and the needs of employees as the life outside of the workplace often influences life in the work atmosphere. NFPA 4.7; 5.7. *CO, 2E,* Page 123 - 124.

Question #51. As the company officer gives orders on the emergency scene, the firefighter, who is the receiver of the message, interprets the message and acts from there. NFPA 4.2.2. *CO, 2E,* Page 21.

Question #52. Individuals in theory with high achievement needs tend to work more diligently and want to see it as an achievement. NFPA 5.2.2. *CO, 2E,* Page 125.

Question #53. The priorities for any incident are as follows in order: life safety, incident stabilization, and property conservation. Operations that make a direct attack on the fire are in the offensive strategy to control the fire and keep it in place. NFPA 4.6.3. *CO, 2E,* Page 327.

Question #54. Fire suppression personnel can effectively inspect many occupancies to detect and even facilitate correcting common fire code violations. NFPA 5.5.1; 4.3.4. *CO, 2E,* Page 226.

Question #55. Whether volunteer or career, the company officer is responsible for the human resources or staffing along with other resources. NFPA 4.2.1; 4.2.2 4.2.3. *CO, 2E,* Page 54.

Question #56. A goal is a target or object by which achievement can be measured. As company officers, we set departmental, company, and individual goals for ourselves and others daily. NFPA 4.2.3. *CO, 2E,* Page 98.

Question #57. You will be a mediator frequently as you try to get others to work out their differences. NFPA 5.7. *CO, 2E,* Page102.

Question #58. Most new employees require a lot of direction and supervision. NFPA 5.2.1, 4.2.1. *CO, 2E,* Page 128.

Question #59. Because this type of construction allows long spans without support, it is very popular in large one-story commercial facilities. NFPA 4.6.2. *CO, 2E,* Page 249.

Question #60. The company officer gives many oral commands on emergency scenes and they follow customs, rules, and practices of the industry, thus making them formal. NFPA 4.2.1. *CO, 2E,* Page 20.

Question #61. All elements of NFPA 1403 and water supply standards must be followed for safety concerns. NFPA 4.7.1. *CO, 2E,* Page 309 & 310.

Question #62. The employer should encourage participation and growth along with providing the environment for this to occur. The individual must take advantage of the opportunities to enhance the skills and knowledge for the advancement. NFPA 4.1.1. *CO, 2E,* Page 13.

Question #63. Most employees described in the question require some discussion, direction, and supervision. They are beginning to contribute ideas and solutions. NFPA 5.2.1, 4.2.1. *CO, 2E,* Page 128.

Question #64. The first stage of the fire is limited to the materials originally ignited. NFPA 4.6.2; 5.6.1. *CO, 2E,* Page 257.

Question #65. As a company officer, it is important to remember to allow in this situation the communications to be a two-way process. NFPA 4.2.2. *CO, 2E,* Page 21.

Question #66. Knowing the leading causes of fire as being careless smoking and the evidence presented, you would tend to lean toward that answer based upon fire behavior and the time of day, coupled with the ashtray finding and the smoking resident. NFPA 4.5.1. *CO, 2E,* Page 221.

Question #67. This is a standardized way of functioning based upon an organized directive that establishes a standard course of action. NFPA 4.6.3; 5.6.1. *CO, 2E,* Page 304.

Question #68. The company officer's job is as varied as all the activities performed by the organization. However, the company officer is the first-line supervisor and generally the senior representative that the public deals with on a routine basis. NFPA 4.1.1. *CO, 2E,* Page 4.

Question #69. It is important to understand the requirements to which you are certifying as a company officer. NFPA 4.1. *CO, 2E,* Page 9.

Question #70. Knowing the leading causes of fire as being careless smoking and the evidence presented, you would tend to lean toward that answer based upon fire behavior and the time of day coupled with the ashtray finding and the smoking resident. However, the accidental causes of fires are lead by heating equipment. The question brings forth temperature, time of year, and potentials. NFPA 4.5.1. *CO, 2E,* Page 273.

Question #71. Most organizations have a pyramid structure, with one person in charge, and an increasing number of subordinates at each level as you move downward. The company officer is the first level of supervisor and has a limited number of subordinates for direct supervision, but numbers increase with the size of station and amount of incidents. NFPA 4.1.1. *CO, 2E,* Page 61.

Question #72. As an organization that has a bargaining union, it is important, as an officer, to understand the factors that surround the contract and the department so as not to violate any laws or conditions that are legally binding. It is important also to know these as you are the representative for your employees' best welfare. NFPA 5.2.1. *CO, 2E,* Page 100.

Question #73. The experience of a fire investigator will prove to be beneficial especially with the elements of potential death and potential arson. NFPA 4.5.1; 4.5.2; 5.5.2. *CO, 2E,* Page 277.

Question #74. The first step in the management process is planning and the effectiveness of a company officer is the ability to make and keep plans moving forward. NFPA 5.4.5. *CO, 2E,* Page 77.

Question #75. Total quality management principle is a style often practiced by many company officers as they focus on the organization's continuous improvements and keeping customer satisfaction in mind. NFPA 5.4.5. *CO, 2E,* Page 81.

Question #76. Understanding an employee's competency and commitment is important as you mentor and work with these individuals as a supervisor. NFPA 4.2.2. *CO, 2E,* Page 146.

Question #77. Understanding an employee's competency and commitment is important as you mentor and work with these individuals as a supervisor. NFPA 4.2.2. *CO, 2E,* Page 146.

Question #78. Groups exist whenever two or more people share a common goal. Organizations are groups of people. Typically, these people share a common goal, have formal rules, and have designated leaders. This is true of fire companies. NFPA 4.2. *CO, 2E,* Page 44.

Question #79. Lightweight construction is engineered to be as strong or stronger than solid components. NFPA 4.6.2. *CO, 2E,* Page 250.

Question #80. This is the document by which you inspect. These codes are adopted by local, state, or other jurisdictional entities. NFPA 5.5.1. *CO, 2E,* Page 228.

Question #81. The inability to say no or decline the opportunity is the hardest of the time eaters for most. You must remember to tie yourself to goals and focus on priorities. It is important to know when you are overloading your plate. Procrastination is just the opposite. You do not have a focus on goals or priorities as it relates to work. The continued putting off of this project has you in a position of crunch time for completion. This usually leads to poor work. NFPA 5.2.1. *CO, 2E,* Page 106.

Question #82. Attitude is the core of your performance. Your attitude when positive will reflect a professional and loyal role model who will be able to lead others to meet the department's mission. NFPA 4.1.1. *CO, 2E,* Page 15.

Question #83. The first step in the management process is planning and the effectiveness of a company officer is the ability to make and keep plans moving forward. NFPA 5.4.5. *CO, 2E,* Page 77.

Question #84. The purpose of the disciplinary action is to improve the subordinate's performance or conduct. NFPA 4.2.5. *CO, 2E,* Page 161.

Question #85. This is one of several leadership tools that focuses the employees toward improving their work performance. NFPA 4.2.5. *CO, 2E,* Page 149.

Question #86. The amount of water that is required theoretically to suppress the fire can be figured by both of these formulas. NFPA 4.6.1. *CO, 2E,* Page 263 & 264.

Question #87. Fire prevention is a proactive approach that is most often accomplished through public contact and educational session done by firefighters. NFPA 4.3.4. *CO, 2E,* Page 214.

Question #88. Length x Width x Height / 100 = the fire flow for each floor. If the floor is 1/4 involved, then divide that answer by 4 and you would get 4,500 gpm. NFPA 4.6.1. *CO, 2E,* Page 264.

Question #89. All high-rise buildings are made of fire-resistive construction. NFPA 4.6.2. *CO, 2E,* Page 248.

Question #90. Action plans address all areas of the fire and in a time sequential order or time frame. NFPA 4.6.3. *CO, 2E,* Page 331.

Question #91. As most fire chiefs would agree they manage primarily from behind a desk, they would also agree that the company officer is a working foreman managing personnel from where the action is occurring just like a floor supervisor in an industrial plant. NFPA 5.6.1; 4.6.3. *CO, 2E,* Page 84.

Question #92. The accomplishment of the organization's goals by utilizing the resources available is management. As a company officer, you manage human resources daily. NFPA 4.2.2. *CO, 2E,* Page 75.

Question #93. It is important to understand the dynamics behind why programs exist within your department and the fire service. As a company officer, you will be faced with questions from the public and youthful firefighters as to why the fire service is engaged in prevention and life safety programs. NFPA 4.3.4. *CO, 2E,* Page 217.

Question #94. A good leader must understand human behavior and the needs of employees as the life outside of the workplace often influences life in the work atmosphere. NFPA 4.7; 5.7. *CO, 2E,* Page 123.

Question #95. Controlling allows us to measure the effectiveness of our effort to help us maintain our goals. By doing so, we can seek ways to improve, thus increasing productivity. NFPA 5.6.1; 5.4.2. *CO, 2E,* Page 79.

Question #96. According to NFPA 1500, every department should have an individual assigned to the duties of health and safety officer. NFPA 5.7.1. *CO, 2E,* Page 184.

Question #97. Each year, over 100 firefighters die in the line of duty. The majority of the fire ground deaths occur when firefighters are advancing hose lines inside the structure. NFPA 4.7.1. *CO, 2E,* Page 197.

Question #98. Regardless of whether we are paid or volunteer, we are serving the public and want to be considered and viewed as professionals. NFPA 4.1.1. *CO, 2E,* Page 8.

Question #99. Company officers, for firefighters, are the first step in the chain of command. Command-level officers consider the company officer to be the lower management staff and the first-line supervisors for the organization. NFPA 4.2. *CO, 2E,* Page 4.

Question #100. With life safety is an incident priority, it is important to understand the pre-planned locations at which these individuals will be located. This will enhance rescue operations. NFPA 4.6.1, 5.5.2. *CO, 2E,* Page 298.

Exam III

1. As the company officer, you average 13 hours of your shift actually working. The 13 hours are broken down 11 hours administrative or non-emergency functioning and 2 hours of emergency response. These are the normal percentages for most company officers.
   a. True
   b. False
2. Career development for a company officer, who is also a firefighter, is a shared responsibility. Both the company officer and the department have an obligation to this growth.
   a. True
   b. False
3. Effective company officers do not have the need to write incident reports effectively.
   a. True
   b. False
4. Strong leadership and closed communications by company officers are vital to the success of any organization.
   a. True
   b. False
5. Good time management correlates the management of our time with our previously set goals and objectives.
   a. True
   b. False
6. You are a 28-year-old engineer who has just been promoted to captain. It is safe to say that you have enjoyed being in an acting officer role for some time. Is the following statement true: Individuals with high power needs like to be in charge?
   a. True
   b. False
7. Persons with high affiliation needs don't like to work in groups.
   a. True
   b. False
8. The cycle of performance management represents the continuous process of goal setting, observation, and performance evaluation.
   a. True
   b. False

9. Nearly half of firefighters killed during structural firefighter operations were advancing hose lines at the time of their injury.
   a. True
   b. False
10. Lightweight truss construction is primarily used in Type I construction.
    a. True
    b. False
11. Good communication skills are essential in both your work and personal life.
    a. True
    b. False
12. Effective fire officers do not have the need to write effectively.
    a. True
    b. False
13. Strong leadership and closed communications are vital to the success of any organization.
    a. True
    b. False
14. You are the company officer of an employee who has called in sick and then attended a live fire training exercise with his part-time department. As the company officer you do not report your findings to the battalion chief. This is a direct violation of the policies. In this case, ethics are not an important management issue.
    a. True
    b. False
15. A fire department's Fire Suppression Rating Schedule is directly related to the rating classification given by the Insurance Service Office.
    a. True
    b. False
16. Planning does not have to start with a clear understanding of the goals and objectives.
    a. True
    b. False
17. Persons with high achievement needs like challenges.
    a. True
    b. False
18. Regardless of whether the firefighter is a career employee or a volunteer member, injuries cost money.
    a. True
    b. False

19. Nearly half of those firefighters killed during structural firefighter operations were advancing hose lines at the time of their injury.

    a. True

    b. False

20. You find a body that has gun shots to the head in a bathtub during a fire. In the investigation, there are poor patterns throughout the house. The crime of arson requires criminal intent as well as the act itself. Is this statement fitting to the scenario described?

    a. True

    b. False

21. You are the company officer giving orders on a fire scene. In the elements of communications process, the company officer would be which of the following elements?

    a. feedback

    b. sender

    c. receiver

    d. medium

22. You are the company officer giving orders on a fire scene. These orders in the communications model would be known as is known as the _______________.

    a. message

    b. feedback

    c. medium

    d. none of the above

23. As a company officer dealing with your duties as an officer in your daily activities, you may experience which of the following as an obstacle in the communications process?

    a. physical barrier

    b. personal barrier

    c. semantic barrier

    d. all of the above

24. You are the company officer of an employee who has called in sick and then attended a live fire training exercise with his part-time department. As the company officer, you do not report your findings to the battalion chief. The battalion chief disciplines you because the act of being accountable for actions and activities along with having a moral and perhaps legal obligation to carry out certain activities is your _______________.

    a. accountability

    b. responsibility

    c. line of authority

    d. all of the above

25. You give assignments at the beginning of the shift to your crew of six that must be completed by 5PM in order to meet a department deadline. This is known as _______________.
    a. advocation
    b. delegation
    c. administration
    d. relocation
26. You are responsible for a live fire training exercise and you assign responsibilities to other officers and run the exercise as they teach specific disciplines. This step in the management process which involves the manager's directing and overseeing the efforts of others is _______________.
    a. coordinating
    b. commanding
    c. organizing
    d. controlling
27. You are hired as an outside individual to fill a vacant captain's role. You are required to become a union member as a condition of your employment. A labor relations term denoting this concept is _______________ shop.
    a. union
    b. closed
    c. open
    d. agency
28. As the captain of a fire station the fire chief gives you a budget to acquire the goods and services to run that particular station for six months. This is which of the following budgets?
    a. capital budget
    b. operating budget
    c. line-item budget
    d. program budget
29. The five-tiered representation of human needs developed by Abraham Maslow is known as Maslow's _______________.
    a. Sequence of Needs
    b. Directional Needs
    c. Hierarchy of Needs
    d. Hierarchy of Desires
30. You are promoted to lieutenant and assigned to a truck company. The members of this new company addresses you by the title Lieutenant Mann. This recognition of authority derived from the government or other appointing agency is _______________ power.
    a. demanded
    b. granted
    c. legitimate
    d. reward

31. The crew you are supervising has over the last two cycles moved an employee's bed outside, short sheeted his bunk, frozen his keys in a glass of water, and put his car up on blocks. This is best described as which of the following?
    a. bug
    b. aggravate
    c. harassment
    d. none of the above

32. You have an employee who has called in sick yet gone on a fishing trip that day with his father. He made attempts to secure an employee trade of duties. He failed to secure this. You suspend the employee for 72 hours without pay. The purpose of this disciplinary process is to improve the subordinate's _______________.
    a. performance
    b. conduct
    c. both a & b
    d. none of the above

33. A captain assigned as the manager of the department's health and safety program is the _______________ officer.
    a. safety
    b. health
    c. health and safety
    d. division of life safety

34. You as a company officer are required to investigate fires to which you respond. Which of the following is the leading cause of civilian fire deaths?
    a. cooking
    b. heating appliances
    c. children
    d. careless smoking

35. A Wal-Mart Super Center store built in 2004 would typically have which of the following building construction types?
    a. Type I
    b. Type II
    c. Type III
    d. Type IV

36. The successful completion of an accredited Fire Officer I course attests that one has demonstrated the knowledge and skills necessary to function in a particular craft or trade is known as _______________.
    a. qualification
    b. training
    c. certification
    d. all of the above

37. The final step in the management process involves monitoring the process to ensure that the work is accomplishing the intended goals and objectives and taking corrective action when it is not. This is known as _______________.
    a. coordinating
    b. organizing
    c. controlling
    d. commanding

38. You give three months for a list of preplans to be completed. The crews turn them in the last week prior to the due date. This is known as _______________.
    a. Managerial Law
    b. Parkinson's Law
    c. Ranle Law
    d. Empowering Law

39. To disturb, torment or pester is _______________.
    a. an officer's responsibility
    b. an officer's right
    c. harassment
    d. fun

40. Your department is made up of 20 percent Hispanic, 25 percent black, 10 percent Native American, and 5 percent females. This mix is known as _______________.
    a. strange
    b. weird
    c. civil
    d. diversity

41. A company officer who helps a team member or subordinate improve their knowledge, skills, and abilities is known as a _______________.
    a. counselor
    b. coach
    c. director
    d. chief

42. In 2002, the leading cause of death to firefighters was _______________.
    a. responding/returning from alarms
    b. training
    c. fire ground
    d. non-fire emergencies

43. You are in Organizational Theory class at the National Fire Academy. They require you to read America Burning. When was it published?
    a. 1903
    b. 1943
    c. 1973
    d. 1993

44. You respond to a reported structure fire with smoke showing and water running from the building. After confirming the fire was out and doing overhaul, you examine the fire suppression system and find a self-operating thermosensitive device that released a spray of water over a designed area and control and eventually extinguished this fire. This is known as a ______________.
    a. deluge gun
    b. stand pipe
    c. prepiped system
    d. automatic fire protection sprinkler

45. You role up to see fire showing from one window on the third floor. You would say the fire is in which of the following phases?
    a. incipient phase
    b. free-burning phase
    c. backdraft phase
    d. smoldering phase

46. The NFPA standard for Firefighter Professional Qualifications is ______________.
    a. 1001
    b. 1002
    c. 1021
    d. 1710

47. As a newly promoted lieutenant, you are ______________.
    a. responsible for the performance of the assigned crew
    b. responsible for the safety of the assigned crew
    c. second-line supervisor
    d. a & b
    e. all of the above

48. As this newly promoted lieutenant, you have assigned personnel. These personnel are dependent upon you for guidance daily. This makes your primary role ______________.
    a. leadership
    b. coaching
    c. mentoring
    d. friendship

49. The term company is used to describe work teams. The individual of that work team who must coordinate activities is the company officer. This individual has many roles to fill. Which of the following is not one of those roles?
    a. help in advancing hose
    b. directing crew operations
    c. activities that don't compromise roles as leaders
    d. developing a mission statement

50. Career development is a shared responsibility between which of the following?
    a. the individual and the community
    b. the individual and the department
    c. the community and the department
    d. the environment and you

51. You are a newly promoted lieutenant in your department of 500 personnel. Your supervisor is a district chief. Where do you fit into the department's organization?
    a. middle management
    b. lower management
    c. first-line supervisor
    d. a & c
    e. b & c

52. During emergency scene operations, the firefighter typically would be the _______ in the communications model.
    a. sender
    b. message
    c. receiver
    d. all of the above

53. You are reviewing an employee's performance evaluation for the quarter. It is important to have good communications occurring. This two-way communications allows the employee to provide ___ on the evaluation.
    a. feedback
    b. knowledge
    c. barriers
    d. all of the above

54. The ABC fire department has a pyramid organizational structure. Where does the company officer fall within this structure?
    a. front-line supervisor
    b. first-line supervisor
    c. second-level supervisor
    d. a & b
    e. a & c

55. ABC fire department has a progression or career development path for individuals to move up through the ranks from firefighter to chief. This uninterrupted series of steps or layers is an organizational concept known as _______________.

a. flat principle
b. scaler principle
c. unity of command
d. division of labor

56. As a captain, you are delegated a project to develop a puppet show stage for fire prevention week. You are to make a prop that resembles a fire truck for a puppet show. You utilize one of your crew member's talent in carpentry. This concept is known as _______________.

a. delegation
b. advocation
c. management
d. supervision

57. You are assigned as the leader of a task force that will be working in a collapsed building. You have selected team members that are specifically trained in the area of structural collapse with specialties in shoring, engineering design, and fiber optic visual technology. As the task force leader, you would be considered to be___________ this group as you begin rescue operations for trapped victims.

a. organizing
b. commanding
c. developing
d. coordinating

58. You are a lieutenant in a department that has a single station. Your station is a multiple company station and you are a seasoned company officer on a rescue company. Your battalion chief asks you to develop a five year station work plan. One year into this plan, you find that you are behind by 2two months. During the next year, you make up this deficit and are on schedule as planned. This is an example of _______________.

a. controlling
b. commanding
c. organizing
d. developing

59. As a lieutenant, you meet with your crew after a large department officer meeting and set a vision for where the company is going based on the department's mission statement. You set clear concise expectations based upon the guiding principles of the department. In developing this, you are focusing on both the internal and external customer's needs. In doing this, you are practicing what management style?

a. Theory X
b. Theory Z
c. traditional management
d. total quality management

60. As a captain, upon promotion, you are afforded the right and power to command. This right and power by position is known as _______________.
    a. respect
    b. power
    c. managerial legal right
    d. authority

61. As an applicant with a department, your employment is conditional on your joining the union. Failure to join or agreement to join resulting in your employment is an example of a(n) _______________ shop.
    a. open
    b. closed
    c. union
    d. agency

62. You have been a captain for 10 years and have been found to be reliable and dependable to the higher administrative staff on smaller assigned projects. The Chief has asked you to head up a hiring process for new recruits. It is summer time and you have a new motor home you and your family are enjoying everyday you are off. You also have taken several days off to go to the lake. You were assigned the hiring project on June 9. The completion date has been set for July 23. Today it is July 13 and you have done very little work on this project. Which one of the following best describes you in the proven time eaters.
    a. lack of personal goals
    b. reacting to urgent events
    c. procrastination
    d. trying to do too much yourself

63. You have a female employee who has just gone through the death of their spouse leaving them with two children and a large sum of bills due to a long illness. This employee has spent countless hours caring for their spouse and continuing to work and care for the children. Which of the following best describes the position into which this person would fall in Maslow's hierarchy of needs?
    a. self-actualization
    b. self-esteem
    c. security needs
    d. social needs

64. Lieutenants have status and power. We add this concept to our list of motivators that get employees to do what is asked of them. You ask a firefighter to take out the trash. The task is quickly completed. One reason would be that you as the company officer have _______________ power.
    a. legitimate
    b. reward
    c. punishment
    d. identification

65. As the company officer, you will use many leadership styles to accomplish the mission of the department and the goals of the company. You have an employee who is a twenty-year veteran and has demonstrated progressing strength in skills and abilities on the job of a five-year veteran. Which leadership style will you most likely utilize?

  a. directing
  b. consulting
  c. supporting
  d. delegating

66. As the company officer, you will use many leadership styles to accomplish the mission of the department and the goals of the company. You have an employee who is a ten-year veteran and has demonstrated strength in skills and abilities on the job. This employee functions at the next level consistently and routinely takes on projects and requires little support from you. Which leadership style will you most likely utilize?

  a. directing
  b. consulting
  c. supporting
  d. delegating

67. You are a well-respected company officer. You are known as a person who helps a team member or subordinate improve their knowledge, skills, and abilities. This is known as a _______________.

  a. counselor
  b. coach
  c. director
  d. chief

68. As a company officer, you have an individual which has been through several significant calls involving death on all three calls. Which of the following are not generally accepted tools for dealing with critical incident stress?

  a. training
  b. debriefing
  c. alcohol
  d. counseling
  e. scene management

69. You pull up on a working two room fire and deploy hose and get a quick knock down on the fire. You crew quickly checks for extension and confirms that all of the fire is out. This is known as fire _______________.

  a. suppression
  b. prevention
  c. attack
  d. department

70. You are in the Inspections course at the National Fire Academy. You are required to read two books, one of which is America Burning Revisited. This book was published in _______________.

   a. 1903
   b. 1943
   c. 1973
   d. 1987

71. You are the lieutenant leading a crew into a heavy smoke condition working fire. As you approach what you believe the seat of the fire to be you see a room on fire that rapidly leads to full involvement of all combustible materials present in the room almost simultaneously. This phenomena is known as _______________.

   a. backdraft
   b. rollover
   c. flashpoint
   d. flashover

72. Lieutenants should remember that they have a right to complete an investigation, and that they should have continuous presence on the property until that investigation is complete or an investigative team has released them. Once the premises is vacated, what has to be secured prior to reentering the premises to do an investigation?

   a. probable cause
   b. search warrant
   c. nothing has to occur
   d. administrative process

73. You are the lieutenant leading a crew into a heavy smoke condition working fire. As you approach what you believe the seat of the fire to be you see a room on fire that rapidly leads to full involvement of all combustible materials present in the room almost simultaneously. You direct your crews to use a straight stream to attack the fire in a combination attack. This is known as a _______________ mode.

   a. defensive
   b. transitional
   c. command
   d. offensive

74. You are directed by the battalion chief, who is the incident commander, for your company to ladder the building on the south side and advance a 2 ½-inch hose line in the second floor to begin an aggressive fire attack. These various maneuvers that can be used to achieve a strategy while fighting a fire or dealing with a similar emergency are known as _______________.

   a. tactics
   b. strategy
   c. size-up
   d. action plan

75. Which of the following is not a building construction type?

    a. ordinary
    b. heavy roof
    c. non-combustible
    d. heavy timber

76. Fire officers must base their actions on priorities such as life safety, property conservation, and incident stabilization. Which of the following puts these in the correct order?

    a. life safety, incident stabilization, property conservation
    b. incident stabilization, life safety, property conservation
    c. property conservation, life safety, incident stabilization
    d. property conservation, incident stabilization, life safety

77. You are the senior captain on scene and have assumed the command role. Your incident is a strip mall fire that has started in the middle of the strip mall and is moving both directions. The strip mall has 25 stores. You direct your crews to cut trench cuts on both sides and set up master streams to control the fire well in front of the fire progression. This act of controlling the fire to prevent the fire from extending any further in the structure is known as _______________.

    a. fire detection
    b. overhaul
    c. extinguishment
    d. confinement

78. You are a lieutenant of a five-man company. As the company officer, you are responsible for the _________ of the assigned personnel.

    a. performance
    b. safety
    c. mission development
    d. both a & b
    e. both a & c

79. You have just spent the past three weekends attending an Inspector II program. Today, you received notification that you passed your state exam and you would receive your paperwork in the mail within five working days. This document that attests you have demonstrated the knowledge and skills necessary to function in a particular craft or trade is known as _______________.

    a. qualification
    b. training
    c. certification
    d. all of the above

80. You have tried to give orders to crews across a busy highway. You are finding that it is hard to communicate due to traffic noise. This noise that is causing problems in communications is known as the _______________.

 a. sender

 b. message

 c. receiver

 d. medium

81. In question 60, this obstacle in communications that prevents the message from being understood by the receiver is a _______________.

 a. barrier

 b. reading

 c. active listening

 d. none of the above

82. You are having problems with an employee always taking your food. You ask the lieutenant to speak to him about this. He takes you in the office and sits in a chair facing you. He tells the other crew members to hold all his incoming calls. The best way to describe the actions of the Lieutenant is known as _______________.

 a. listening

 b. active listening

 c. passive listening

 d. aggressive listening

83. You have been from firefighter to master firefighter followed by a promotion to lieutenant and now you are being promoted to captain. This progression of an uninterrupted series of steps or layers within an organization refers to which of the following?

 a. linear principle

 b. unity principle

 c. scaler principle

 d. division of labor principle

84. Which of the following is the first step in the management process and involves looking to the future and determining objectives?

 a. organizing

 b. planning

 c. commanding

 d. coordinating

85. You are asked to chair an apparatus specifications committee. You have worked on the specifications for the truck and have determined it will cost $1,000,000 to replace Tower 71 which is a 105-foot aerial platform apparatus. You would put this cost in which of the following budgets?

    a. line-item budget
    b. program budget
    c. capital budget
    d. operating budget

86. You are budgeting for training tapes and AV equipment for the training budget. Collecting these similar items into a single account and presenting them in a general area collectively under this topic or one line in a budget document is _______________ budget.

    a. capital
    b. operating
    c. program
    d. line-item

87. Which of the following is not one of the three major components of customer service?

    a. Constantly be selective of the services.
    b. Always be nice.
    c. Regard everyone as a customer.
    d. Constantly raise the bar.

88. You give an assignment of preplan to a company and set a deadline of three months to complete. They complete the list the final week. The following year, you give them the same list and set the deadline at one month. They complete the project on the last day prior to the deadline. This time use is known as the _______________.

    a. Managerial Law
    b. Parkinson's Law
    c. Range Law
    d. Empowering Law

89. As a lieutenant on an engine company, in which of the following three essential elements in fire prevention would you be the least involved?

    a. education
    b. engineering
    c. enforcement

90. As a lieutenant on an engine company, in which of the following three essential elements in fire prevention would you be the most involved?

    a. education
    b. engineering
    c. enforcement

91. You are the senior lieutenant who has been tasked at looking into the health and safety program of the department. You are tasked with bringing the department's rules, regulations, and standard operating guidelines into compliance with professional standards and laws. Which two organizations would you look to for information and guidance?
    a. JEMS and FIREHOUSE Magazines
    b. NIOSH and JOCCA
    c. NFPA and JOCCA
    d. NFPA and OSHA

92. You have been certified as a Level II Inspector and officially sworn in. You are conducting company fire inspections in your district. The reference book you are using is based on the BOCCA. As a company officer, you are issued a legal document that sets forth the requirements for life safety and property protection in the event of fire, explosion, or similar emergency. This legal document is which of the following?
    a. fire prevention code
    b. building code
    c. minimal standards
    d. elements and standards

93. You have James Poe as your firefighter. He has been with the department four years and has been released to function as a relief driver and as a relief company officer. He is in charge of the FIREHOUSE software program for the entire department. He continuously looks for ways to improve himself and is constantly taking weekend schools when he is off. He just completed his instructor certification for Live Fire Training. You would say this person is which of the following?
    a. high-achievement needs
    b. high-affiliation needs
    c. low-achievement needs
    d. low-affiliation needs

94. The addition of an automatic wet sprinkler system to a building lowers which of the following?
    a. life risk factors
    b. property risk factors
    c. physical risk factors
    d. community risk factors

95. As a lieutenant and a life safety educator, you are always looking for the advantage over emergency responses, especially fire. Which one of the following would be classified as one of our greatest allies in fire protection?
    a. construction features
    b. fire codes
    c. sprinkler systems
    d. smoke detectors

96. You are the investigating officer on a small room and contents fire in a bedroom. The fire began about 7 AM on Friday, February 1. The temperature outside is -10 degrees and there is 8 inches of snow on the ground. You have conducted interviews with the occupants who are smokers and are now looking at findings in the room. The bed is burned significantly and the heat line of demarcation is about 5 feet off the floor. There is heavy smoke residue in the room with minor fire damage to any other furniture.You find a space heater plugged in up close to the window beside the bed, several extension cords, and an ashtray by the bed. What would you suspect the cause of this fire to be?

    a. electrical
    b. improperly positioned space heater
    c. smoking
    d. not enough information to determine

97. You are the senior company officer and the first arriving company officer to the scene of a multi-family residential apartment complex on Thanksgiving night. You arrive to find this 10-story high-rise apartment complex with heavy fire involvement on the first and second floors of one end involving at least four apartments. Dispatch had a subject on the phone begging for help when the phone was dropped and the line remained open for a period of time before going dead. You now have brought the fire under control. You have a victim unaccounted for at this time. Which of the following meets the guidelines for calling a fire investigator?

    a. potentially incendiary fire
    b. a fire resulting in a death
    c. property damage greater than $50,000
    d. a & b
    e. all of the above

98. You have a large oil refinery and tank farm distribution center in your district. This type of complex is an example of a _______________.

    a. fire target
    b. target hazard
    c. facility with multiple automatic alarms daily
    d. none of the above

99. Use the Iowa State formula. You respond to a commercial structure which is 150 feet wide, 500 feet long, and 12 feet high on each of 50 floors. With this knowledge, figure fire flow based on an appropriate theoretical fire flow formula utilizing all of the building's dimensions. How much would the fire flow be (in gpms) if the building was 25% involved on the 50th floor?

    a. 10, 000 gpm
    b. 2,250 gpm
    c. 3,000 gpm
    d. 30,000 gpm

100. The difference between the fire triangle and the fire tetrahedron is _______________.

a. heat

b. fuel

c. chemical chain reaction

d. oxygen

Phase III, Exam III: Answers to Questions

1.	T	26.	A	51.	E	76.	A
2.	T	27.	B	52.	C	77.	D
3.	F	28.	B	53.	A	78.	D
4.	F	29.	C	54.	D	79.	C
5.	T	30.	C	55.	B	80.	D
6.	T	31.	C	56.	C	81.	A
7.	F	32.	A	57.	B	82.	B
8.	T	33.	C	58.	A	83.	C
9.	T	34.	D	59.	D	84.	B
10.	F	35.	B	60.	D	85.	C
11.	T	36.	C	61.	B	86.	D
12.	F	37.	C	62.	C	87.	A
13.	F	38.	B	63.	C	88.	B
14.	F	39.	C	64.	C	89.	B
15.	T	40.	D	65.	B	90.	A
16.	F	41.	B	66.	D	91.	D
17.	T	42.	C	67.	B	92.	A
18.	T	43.	C	68.	C	93.	A
19.	T	44.	D	69.	A	94.	A
20.	T	45.	B	70.	D	95.	C
21.	B	46.	A	71.	D	96.	B
22.	A	47.	D	72.	B	97.	E
23.	D	48.	A	73.	D	98.	B
24.	B	49.	D	74.	A	99.	B
25.	B	50.	B	75.	B	100.	C

Phase III, Exam III: Rationale & References for Questions

Question #1. Most of the time officers deal with other functions rather than emergency response. NFPA 1021 4.2. *CO, 2E,* Page 5.

Question #2. The officer helps the firefighter develop based on the community needs. NFPA 2-2.1. *CO, 2E,* Page 13.

Question #3. Being able to write is an important part of your professional career. NFPA 4.1.2. *CO, 2E,* Page 28.

Question #4. Most structured departments have a pyramid-style design that it is important to be able to communicate from top to bottom and bottom to top. NFPA 4.2. *CO, 2E,* Page 60.

Question #5. It is important to be able to manage time to have good efficiency. NFPA 4.7 5.7.1. *CO, 2E,* Page 105.

Question #6. Most officers have high-power needs. NFPA 5.2. *CO, 2E,* Page 125.

Question #7. The fire service is typically a team organization in every aspect. NFPA 4.2.2, 5.2. *CO, 2E,* Page 125.

Question #8. This is a cycle due to the achievements and progressional growth of personnel. NFPA 5.2.2. *CO, 2E,* Page 152.

Question #9. This is due to the fact most firefighters die early on in incidents and hose advancement is generally in early stages of incidents. NFPA 4.7.2, 5.7.1. *CO, 2E,* Page 178.

Question #10. Lightweight construction is in Type II or Type V based on if it is wood or steel. NFPA 4.6.1. *CO, 2E,* Page 251.

Question #11. Effective communications are a vital part of your professional life. NFPA 4.1.2. *CO, 2E,* Page 20.

Question #12. Being able to write is an important part of your professional career. NFPA 4.1.2. *CO, 2E,* Page 28.

Question #13. Most structured departments have a pyramid-style design that it is important to be able to communicate from top to bottom and bottom to top. NFPA 4.2. *CO, 2E,* Page 60.

Question #14. Every person in the fire service should live by the fire service code of ethics. NFPA 4.3.2 and 4.3.3. *CO, 2E,* Page 91.

Question #15. The fire officer helps meet all of the objectives and the grading of ISO. NFPA 5.1.1. *CO, 2E,* Page 152 on manuscript.

Question #16. Planning is a systematic process and should be on-going. NFPA 5.2.1, 4.2.6, 4.2.1. *CO, 2E,* Page 99.

Question #17. Most officers enjoy the challenges with which they are faced because to get to the officer level, you must be a high achiever. NFPA 5.2, 4.2.2. *CO, 2E,* Page 125.

Question #18. Each workers compensation claim is money paid out. NFPA 5.4.2, 5.7.1. *CO, 2E,* Page 177.

Question #19. This is due to the fact most firefighters die early on in incidents and hose advancement is generally in early stages of incidents. NFPA 4.7.2, 5.7.1. *CO, 2E,* Page 179.

Question #20. Arson has to be targeted at someone. NFPA 4.5.1, 4.5.2, 5.5.2. *CO, 2E,* Page 271.

Question #21. The sender must formulate a thought and send it out to a receiver. NFPA 4.1.2. *CO, 2E,* Page 21.

Question #22. Good communications is essential in both work and personal life. NFPA 4.1.2. *CO, 2E,* Page 21.

Question #23. Barriers can obscure messages. NFPA 4.2.1 section A. *CO, 2E,* Page 23 & 24.

Question #24. We as officers must take responsibility for our actions to be responsible for the actions of others. NFPA 4.2.6 section A. *CO, 2E,* Page 62.

Question #25. To accomplish the goals of the organization, it is important to utilize all of your resources. NFPA 4.6.3. *CO, 2E,* Page 67.

Question #26. To make a gear system run, it must be synchronized. The controlling of personnel efforts is like a machine running with the employees being the gears. NFPA 4.4.2 4.6.2. *CO, 2E,* Page 78.

Question #27. Working in a union environment is a very different situation and will vary from union to union. It is important to understand how the union functions and its goals. NFPA 4.2, 5.2.1, 5.1.1. *CO, 2E,* Page 102.

Question #28. Most officers are aware of the operating budget since this is the concept we use most in our personal lives. NFPA 4.4.3, 5.4.2. *CO, 2E,* Page 103.

Question #29. Human behavior is important in the leadership and management styles you choose as an officer. NFPA 4.2.4, 5.2.1. *CO, 2E,* Page 123.

Question #30. We are given this legitimate power when we are promoted to the officer level by the governing body. NFPA 4.2, 5.2. *CO, 2E,* Page 127.

Question #31. This is an EEOC directive. NFPA 5.2.1; 4.2.1. *CO, 2E,* Page 133.

Question #32. You will be required to take disciplinary actions as an officer. It is important to use them as a tool for redirection of personnel onto the department's mission pathway. NFPA 5.2.1. *CO, 2E,* Page160.

Question #33. This is a standard set out by NFPA 1500. NFPA 5.7, 4.7. *CO, 2E,* Page 184.

Question #34. As officers, it is important to understand the causes of fires and fire fatalities so you can reinforce the materials delivered during fire safety programs and contacts. NFPA 4.5.1. *CO, 2E,* Page 222.

Question #35. Building construction and fire behavior go hand in hand. It is important to understand how fire will behave in particular build construction styles. NFPA 4.6.1. *CO, 2E,* Page 248 & 249.

Question #36. Certification means that an individual has been tested by an accredited examining body on clearly identified material and found to meet the minimum standard. NFPA 3.3.11. *CO, 2E,* Page 8.

Question #37. Controlling helps us get to the right place at the right time through the monitoring of efforts of the resources. This is a large part of the company officer's job. NFPA 4.2.1. *CO, 2E,* Page 79.

Question #38. Managing time is an important aspect of a company officer. Time is the one thing that is hard to track for efficiency based upon the tasks performed and the personnel performing them. It is important, however, not to fall into this concept of allowing the prescribed time to dictate the time needed for the event. NFPA5.2.1; 5.2.2; 5.2.3; 4.4.2. *CO, 2E,* Page 1058.

Question #39. Harassment of any kind should not be allowed in the workplace. Title VII of the Civil Rights Act of 1964 NFPA 4.7.1. *CO, 2E,* Page 133.

Question #40. To have our firefighter population meet the needs of the citizens, it is important to have the total community in view. This means to have representation of various cultures working to help create better understanding and service needs to the entire community. NFPA 4.7.1. *CO, 2E,* Page 132.

Question #41. Coaching is an informational process that helps subordinates improve their skills and abilities. Coaching implies one-on-one relationships that treat subordinates as full partners. NFPA 5.2.1. *CO, 2E,* Page 147.

Question #42. It is important to understand the causes of injuries and death of firefighters to take a proactive approach to reduce these numbers. NFPA 4.7.1; 4.7.2. *CO, 2E,* Page 179.

Question #43. NFPA 4.7. *CO, 2E,* Page 214.

Question #44. NFPA 4.6.1, 5.6.1. *CO, 2E,* Page 254.

Question #45. NFPA 4.6.2, 5.6.2. *CO, 2E,* Page 257.

Question #46. This standard provides the basic requirements for firefighters in their proficiency at the job requirements. NFPA 4.7.1. *CO, 2E,* Page 308.

Question #47. The company officer's job is as varied as all the activities performed by the organization. However, the company officer is the first-line supervisor and generally the senior representative that the public deals with on a routine basis. NFPA 4.1.1. *CO, 2E,* Page 4.

Question #48. Leading others is the company officer's primary job as you are the one who is assigned to direct the personnel assigned to you. To accomplish the mission, you must lead others. NFPA 4.2.1. *CO, 2E,* Page 5.

Question #49. Company officers are expected to lend a hand when needed, whether it is advancing hose or other emergency scene operations so long as it doesn't compromise role as a leader. The company officer is not setting the mission or developing it. They are the ones working with the crews to meet or accomplish it. NFPA 4.2.1. *CO, 2E,* Page 4-6.

Question #50. The employer should encourage participation and growth along with provide the environment for this to occur. The individual must take advantage of the opportunities to enhance the skills and knowledge for the advancement. NFPA 4.1.1. *CO, 2E,* Page 12.

Question #51. Company officers, for firefighters, are the first step in the chain of command. Command-level officers consider the company officer to be the lower management staff and the first-line supervisors for the organization. NFPA 4.2. *CO, 2E,* Page 4.

Question #52. As the company officer gives orders on the emergency scene, the firefighter, who is the receiver of the message, interprets the message and acts from there. NFPA 4.2.2. *CO, 2E,* Page 21.

Question #53. As a company officer, it is important to remember to allow, in this situation, the communications to be a two-way process. NFPA 4.2.2. *CO, 2E,* Page 21.

Question #54. Most organizations have a pyramid structure, with one person in charge, and an increasing number of subordinates at each level as you move downward. The company officer is the first level of supervisor and has a limited number of subordinates for direct supervision, but numbers increase with the size of station and number of incidents. NFPA 4.1.1. *CO, 2E,* Page 62.

Question #55. Principle of scaler organizations is that as you progress with experience and talents, you move upward to higher levels. This is important whereas the company officer is the first level at which individuals accept responsibility for themselves and others. NFPA 4.2. *CO, 2E,* Page 66.

Question #56. The accomplishment of the organization's goals by utilizing the resources available is management. As a company officer, you manage human resources daily. NFPA 4.2.2. *CO, 2E,* Page 75.

Question #57. The third step in the management process, commanding, involves using the talents of others, giving them directions, and setting them to work. NFPA 5.6.1. *CO, 2E,* Page 78.

Question #58. Controlling allows us to measure the effectiveness of our effort to help us maintain our goals. By doing so, we can seek ways to improve, thus increasing productivity. NFPA 5.6.1; 5.4.2. *CO, 2E,* Page 79.

Question #59. Total quality management principle is a style often practiced by many company officers as they focus on the organization's continuous improvements and keeping customer satisfaction in mind. NFPA 5.4.5. *CO, 2E,* Page 82.

Question #60. The position of rank affords the right and the power to command. This is the authority you are granted as a company officer. NFPA 4.4.1. *CO, 2E,* Page 99.

Question #61. As an organization that has a bargaining union, it is important, as an officer, to understand the factors that surround the contract and the department so as not to violate any laws or conditions that are legally binding. It is important also to know these as you are the representative for your employee's best welfare. NFPA 5.2.1. *CO, 2E,* Page 102.

Question #62. The inability to say no or decline the opportunity is the hardest of the time eaters for most. You must remember to tie yourself to goals and focus on priorities. It is important to know when you are overloading your plate. Procrastination is just the opposite. You do not have a focus on goals or priorities as it relates to work. The continued putting off of this project has you in a position of crunch time for completion. This usually leads to poor work. NFPA 5.2.1. *CO, 2E,* Page 106.

Question #63. A good leader must understand human behavior and the needs of employees as the life outside of the workplace often influences life in the work atmosphere. NFPA 4.7; 5.7. *CO, 2E,* Page 123.

Question #64. The recognition of the power of position accompanied by the authority by virtue of the supervisor's ability to administer punishment leads others to see the punishment power concept. NFPA 5.2.1. *CO, 2E,* Page127.

Question #65. Most employees described in the question require some discussion, direction, and supervision. They are beginning to contribute ideas and solutions. NFPA 5.2.1, 4.2.1. *CO, 2E,* Page 128.

Question #66. Most employees described in the question require little discussion, direction, and supervision. They are at the next level waiting for challenges and opportunities to present themselves. NFPA 5.2.1, 4.2.1. *CO, 2E,* Page 128.

Question #67. Coaching is an informational process that helps subordinates improve their skills and abilities. Coaching implies one-on-one relationship that treats subordinates as full partners. NFPA 5.2.1. *CO, 2E,* Page 147.

Question #68. Critical incident stress is a significant part of an emergency responder's life. Assistance is often necessary. NFPA 4.6.4. *CO, 2E,* Page 206.

Question #69. Fire suppression is a reaction. Resources are mobilized following an event and the action to mitigate an event. NFPA 4.6.3. *CO, 2E,* Page 214.

Question #70. NFPA 4.7. *CO, 2E,* Page 214.

Question #71. Flashover is a fire phenomena that presents significant risk to firefighters. Fire behavior is a crucial link to how we formulate tactics to accomplish our strategic goals. NFPA 4.7.1, 4.6.2, 5.6.2. *CO, 2E,* Page 258.

Question #72. Fire investigation is a part of legal proceedings that must follow set steps in order for the materials and information gathered to be used in a legal process. NFPA 4.5.1, 4.5.2, 5.5.2. *CO, 2E,* Page 281.

Question #73. During an offensive mode, operations are conducted in an aggressive manor. NFPA 4.6 , 5.6. *CO, 2E,* Page 327.

Question #74. These operations must be focused on a specific location and measurable as to their success. NFPA 4.6 , 5.6. *CO, 2E,* Page 333.

Question #75. Building construction will define fire behavior and the tactics required to extinguish a fire. NFPA 4.6.2. *CO, 2E,* Page 248-251.

Question #76. The incident commander and the company officer are both responsible for these items, whereas the initial arriving company officer is the incident commander for at least a few minutes. These priorities are how action plans are formulated. NFPA 5.6.1. *CO, 2E,* Page 320.

Question #77. Benchmarks or progression in fire attack are important to the company officer as definitive positions in the incident. NFPA 4.6.3. *CO, 2E,* Page 335.

Question #78. The role of the company officer is to manage personnel at the crew level. Performance and safety are key items for which the officer is responsible. NFPA 4.2.1. *CO, 2E,* Page 4.

Question #79. Certification means that an individual has been tested by an accredited examining body on clearly identified material and was found to meet the minimum standard. NFPA 3.3.11. *CO, 2E,* Page 8.

Question #80. Mediums can have a variety of problems including noise, language, terminology, and others. It is important to try to control this area as much as possible so your communications will be as effective and as clear as possible. NFPA 4.3.3; 4.2.1; 5.2.1; 5.2.2. *CO, 2E,* Page 21.

Question #81. Consider a barrier to be like a filter. One or more filters reduces the information flow between the sender and the receiver. NFPA 4.2; 5.2. *CO, 2E,* Page 23.

Question #82. In active listening, the listener can actually show the sender that you are actively listening by focusing all of your attention on the speaker and showing a genuine interest in the message. Active listening is a significant portion of good communications. NFPA 5.2 : 4.2. *CO, 2E,* Page 24.

Question #83. Like playing a scale on a musical instrument where every note is sounded, the scaler principle suggests that every level in the organization is considered in the flow of communications. NFPA 4.4.2. *CO, 2E,* Page 66.

Question #84. Planning covers everything from the next hour to the next decade. NFPA 4.2.3. *CO, 2E,* Page 77.

Question #85. The operating budget is a general budget that shows specific amounts needed to operate the organization in general form. All large-ticket items of generally one-time frequency until the replacement cycle returns is known as the capital budget. NFPA 5.4.2; 4.4.3. *CO, 2E,* Page 103.

Question #86. Line item budgeting is like keeping accounting on the money spent. This is used when you need specifics to be purchased and don't want it to be generic in nature. NFPA 4.4.3; 5.4.2. *CO, 2E,* Page 103.

Question #87. NFPA 4.3.1; 4.3.3. *CO, 2E,* Page 110.

Question #88. Managing time is an important aspect of a company officer. Time is the one thing that is hard to track for efficiency based upon the tasks performed and the personnel performing them. It is important, however, not to fall into this concept of allowing the prescribed time to dictate the time needed for the event. NFPA 5.2.1; 5.2.2; 5.2.3; 4.4.2. *CO, 2E,* Page 105.

Question #89. This area is generally left to fire protection engineers and building code officials to assure they meet standards and will protect the structure. NFPA 5.5.1; 4.3.4. *CO, 2E,* Page 223.

Question #90. This area is generally left to fire protection engineers and building code officials to assure they meet standards and will protect the structure. NFPA 5.5.1; 4.3.4. *CO, 2E,* Page 223.

Question #91. Safety comes from the elimination of unsafe acts or equipment. It is important as a company officer to recognize the issues and the requirements and standards that support those issues. NFPA 4.7.1. *CO, 2E,* Page 182-183.

Question #92. This is the document by which you inspect. These codes are adopted by local, state, or other jurisdictional entities. NFPA 5.5.1. *CO, 2E,* Page 228.

Question #93. Individuals in theory with high achievement needs tend to work more diligently and want to see it as an achievement. NFPA 5.2.2. *CO, 2E,* Page 125.

Question #94. Life risk factors are affected by the number of people at risk and their danger and ability to provide for their own safety. There has been no reported fire death in a sprinkled building. NFPA 4.3.1. *CO, 2E,* Page 248.

Question #95. Fire suppression systems like sprinklers were designed to control the fire growth or even suppress the fire. With this assistance, we already have no reported deaths within sprinklered buildings. This system is like giving the fire companies additional personnel for suppression efforts that are housed in each occupancy. NFPA 4.6.2; 4.6.3. *CO, 2E,* Page 253.

Question #96. Knowing the leading causes of fire to be careless smoking and the evidence presented, you would tend to lean toward that answer based upon fire behavior and the time of day coupled with the ashtray finding and the smoking resident. However the accidental causes of fires are lead by heating equipment. The question brings forth temperature, time of year, and potentials. NFPA 4.5.1. *CO, 2E,* Page 221.

Question #97. The experience of a fire investigator will prove to be beneficial especially with the elements of potential death and potential arson. NFPA 4.5.1; 4.5.2; 5.5.2. *CO, 2E,* Page 277.

Question #98. Locations where there are unusual hazards or where an incident would overload the department's resources are examples of target hazards. Nursing facilities and retirement centers are high life hazard complexes. NFPA 4.6.1. *CO, 2E,* Page 293.

Question #99. Length x Width x Height / 100 = the fire flow for each floor. If the floor is 1/4 involved, then divide that answer by 4 and you would get 2,250 gpm. NFPA 4.6.1. *CO, 2E,* Page 264.

Question #100. As a company officer, understanding fire growth is important as you make multiple split-second decisions on tactics and operations based upon the knowledge of fire behavior. NFPA 4.6.2. *CO, 2E,* Page 272.

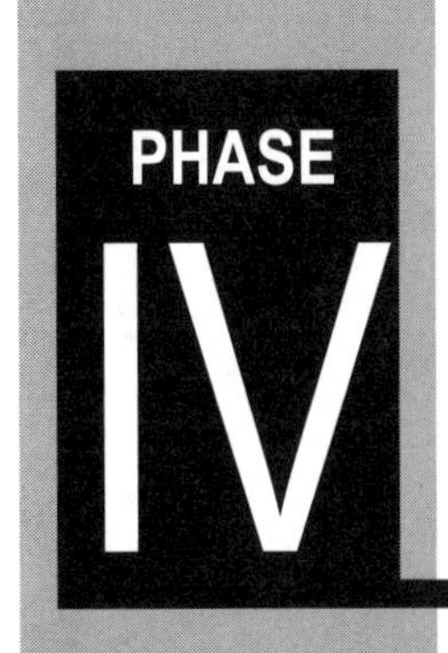

FINAL EXAM

This the final section in the Exam Prep Guide. For this section, we addressed all levels of Bloom's Taxonomy, Cognitive Domain, and all the previous sections. When taking the Final Exam from this section, you will find a variety of questions from basic understanding to application level questions. One should have successfully completed the previous sections before attempting Section Four. Successful completion of this section would indicate a strong knowledge of the material and an in-depth understanding of the content.

1. Regardless of whether the firefighter is a career employee or a volunteer member, injuries cost money.
   a. True
   b. False
2. Good communication skills are essential in both your work and personal life
   a. True
   b. False
3. The cycle of performance management represents the continuous process of goal setting, observation, and performance evaluation.
   a. True
   b. False
4. Persons with high affiliation needs don't like to work in groups.
   a. True
   b. False
5. Nearly half of those firefighters killed during structural firefighter operations were advancing hose lines at the time of their injury.
   a. True
   b. False
6. Good time management correlates the management of our time with our previously set goals and objectives.
   a. True
   b. False
7. Persons with high-achievement needs like challenges.
   a. True
   b. False
8. Persons with high-power needs like to be in charge.
   a. True
   b. False
9. A fire department's Fire Suppression Rating Schedule is directly related to the rating classification given by the Insurance Service Office.
   a. True
   b. False
10. Experienced fire officers should be able to determine the origin and cause of nearly all the fires they attend.
    a. True
    b. False
11. Lightweight truss construction is primarily used in Type I construction.
    a. True
    b. False

12. Career development is a shared responsibility. Both the firefighter and the community have an obligation to this growth.
    a. True
    b. False
13. The crime of arson requires criminal intent as well as the act itself.
    a. True
    b. False
14. In most departments, companies spend less than 10 percent of their time dealing with emergencies.
    a. True
    b. False
15. Ethics are not an important management issue.
    a. True
    b. False
16. Effective fire officers do not have the need to write effectively.
    a. True
    b. False
17. One component in successfully introducing and managing change is to keep communications open.
    a. True
    b. False
18. Planning does not have to start with a clear understanding of the goals and objectives.
    a. True
    b. False
19. Leading others is the company officer's principle job.
    a. True
    b. False
20. Strong leadership and closed communications are vital to the success of any organization.
    a. True
    b. False
21. Locations where there are unusual hazards, where an incident would likely overload the department's resources, or where there is a need for interagency cooperation to mitigate the hazard, is known as _______________.
    a. hazardous material sites
    b. universities
    c. target hazards
    d. none of the above

22. You have been a company officer for five years and have been found to be reliable and dependable to the higher administrative staff on smaller assigned projects. The chief has asked you to head up a hiring process for new recruits. It is summer time and you have a new boat you and your family are enjoying everyday you are off. You also have taken several days off to go to the lake. You were assigned the hiring project on June 9. The completion date has been set for July 23. Today it is July 13 and you have done very little work on this project. Which one of the following best describes you in the proven time eaters?
    a. lack of personal goals
    b. reacting to urgent events
    c. procrastination
    d. trying to do too much yourself
23. You are the company officer of an employee who has thrown a firecracker in the day room where there are several other employees. The firecracker detonated. This caused one of the employees to have some pain in their ear. Which of the following would be the most appropriate answer?
    a. applauding the employee
    b. coaching the employee
    c. counseling of the employee
    d. ignoring the situation; it was a joke
24. Which of the following is a legal document that sets forth requirements to protect health, safety, and the general welfare of the public as they relate to construction and occupancy of buildings and construction?
    a. fire prevention code
    b. building code
    c. city codes
    d. fire codes
25. Which of the following is a part of effective writing skills?
    a. consider the reader
    b. emphasis
    c. brevity
    d. simplicity
    e. all of the above
26. Which of the following is not a building construction type?
    a. ordinary
    b. heavy roof
    c. non-combustible
    d. heavy timber

27. A public fire department is part of which of the following?
    a. community
    b. fire district
    c. organizational development
    d. local government

28. A bird's-eye view of property showing existing structures for purpose of pre-incident planning is known as a ______________ plan.
    a. floor
    b. plot
    c. life safety
    d. map

29. You respond to the older portion of the city for a reported fire in a four-story commercial building. This portion of the city was mostly constructed in the 1920's. Which type of building construction would you anticipate to see?
    a. fire-resistive
    b. non-combustible
    c. heavy timber
    d. ordinary

30. You have an employee who has just gone through the death of their spouse leaving them with two children and a large sum of bills due to a long illness. This employee has spent countless hours caring for their spouse and continuing to work and care for the children. Which of the following best describes the position into which this person would fall in Maslow's hierarchy of needs?
    a. self-actualization
    b. self-esteem
    c. security needs
    d. social needs

31. You respond to a commercial structure which is 150 feet wide, 400 feet long, and 10 feet high. With this knowledge, you could figure fire flow based on which theoretical fire flow formula?
    a. National Fire Academy Formula
    b. Iowa State University Formula
    c. West Virginia University Formula
    d. a & b
    e. a & c

32. Safety is whose concern?
    a. company officer
    b. everyone
    c. safety officer
    d. fire chief
    e. all of the above

33. Significant points in the emergency event usually mark the accomplishments of one of three incident priorities. This process is called ______________.
    a. benchmarking
    b. time marking
    c. trend marking
    d. time elapsed

34. A volunteer fire department would be which of the following?
    a. formal organization
    b. informal organization
    c. combination organization
    d. all of the above

35. You are the investigating officer on a small room and contents fire in a bedroom. The fire began about 11 PM on Friday, August 21. You have conducted interviews with the occupants who are smokers and are now looking at findings in the room. The bed is burned significantly and the heat line of demarcation is about 5 feet off the floor. There is heavy smoke residue in the room with minor fire damage to any other furniture. You find a space heater plugged in up close to the window beside the bed, several extension cords, and an ashtray by the bed. What would you suspect the cause of this fire to be?
    a. electrical
    b. improperly positioned space heater
    c. smoking
    d. not enough information to determine

36. As a company officer, you are delegated a project to develop a public education prop for fire prevention week. You are to make a prop that resembles a fire truck for a puppet show. You utilize one of your crew member's talent in carpentry. This concept is known as ______________.
    a. delegation
    b. advocation
    c. management
    d. supervision

37. You arrive on scene of a two-story woodframe single residential structure that you were dispatched to for an automatic fire alarm. Upon entering the structure, you find a light haze of smoke on the first floor. Further investigation leads you to a smoldering waste basket in the bathroom. This fire was in which of the following stages?
    a. incipient
    b. free-burning
    c. backdraft
    d. smoldering

38. An action taken to control and extinguish fire is known as fire_______________.

    a. suppression

    b. prevention

    c. attack

    d. department

39. As a company officer, effective communications will _______________. 1. enhance leadership ability 2. gain respect from peers and superiors 3. assist in dealing with the media

    a. 1 & 2

    b. 2 & 3

    c. 1, 2, & 3

    d. 1 only

    e. 3 only

40. The difference between the fire triangle and the fire tetrahedron is _______________.

    a. heat

    b. fuel

    c. chemical chain reaction

    d. oxygen

41. A person who helps a team member or subordinate improve their knowledge, skills, and abilities is known as a _______________.

    a. counselor

    b. coach

    c. director

    d. chief

42. Fire officers must base their actions on of priorities such as life safety, property conservation and incident stabilization. Which of the following puts these in the correct order?

    a. life safety, incident stabilization, property conservation

    b. incident stabilization, life safety, property conservation

    c. property conservation, life safety, incident stabilization

    d. property conservation, incident stabilization, life safety

43. The father of professional management was _______________.

    a. McGregor

    b. Brunacini

    c. Drucker

    d. Fayol

44. You are the senior company officer and the first arriving company officer to the scene of a multi-family residential apartment complex on Halloween night. You arrive to find this two-story apartment complex with heavy fire involvement on the first and second floors of one end involving at least four apartments. Dispatch had a subject on the phone begging for help when the phone was dropped and the line remained open for a period of time before going dead. You now have brought the fire under control. You have a victim unaccounted for at this time. Which of the following meet the guidelines for calling a Fire Investigator?
    a. potentially incendiary fire
    b. a fire resulting in a death
    c. property damage greater than $50,000
    d. a & b
    e. all of the above
45. The area that tends to cause problems in communications is often _______________.
    a. sender
    b. message
    c. receiver
    d. medium
46. Use the Iowa State formula. You respond to a commercial structure which is 150 feet wide, 400 feet long, and 10 feet high on each of 50 floors. With this knowledge, figure fire flow based on an appropriate theoretical fire flow formula utilizing all of the building's dimensions. How much would the fire flow be (in gpms) if the building was 25% involved on the 8th floor?
    a. 10, 000 gpm
    b. 1,500 gpm
    c. 3,000 gpm
    d. 30,000 gpm
47. As a company officer, you will set levels of achievement with your employees that can be effectively measured to assure that achievement has occurred. In setting these levels, you are helping the employee set obtainable _______________.
    a. objectives
    b. JPRs
    c. progress
    d. goals
48. This sets broad goals and outlines the overall plan to control the incident.
    a. tactics
    b. strategy
    c. size-up
    d. action plan

49. Professionalism encompasses all of the following except ______________.
    a. attitude
    b. behavior
    c. cognitive
    d. demeanor
    e. ethics
50. A formal dispute between employee and employers over some condition or conditions of employments is a(n) ______________.
    a. ominous sign
    b. conflict
    c. complaint
    d. grievance
51. If the deputy chief of operations would come by a station to have a friendly lunch with the station personnel this would be an example of a(n) ______________.
    a. informal group
    b. formal group
    c. formal relationship
    d. informal relationship
52. The company officer spends _____ percent of their time dealing with administrative aspects.
    a. 10
    b. 40
    c. 50
    d. 90
53. Career development is a shared responsibility between which of the following?
    a. the individual and the community
    b. the individual and the department
    c. the community and the department
    d. the environment and you
54. Firefighting operations that make a direct attack on the fire for purposes of control and extinguishment is ______________ mode.
    a. defensive
    b. transitional
    c. command
    d. offensive
55. A company in a fire department is which of the following?
    a. formal organization
    b. informal organization
    c. formal organization within an informal organization
    d. all of the above

56. As a company officer, you are issued a legal document that sets forth the requirements for life safety and property protection in the event of fire, explosion, or similar emergency. ITS PURPOSE IS TO MINIMIZE THE RISK OF LOSS OF LIFE AND PROPERTY BY REGULATING THE USE AND STORAGE OF MATERIALS THAT MIGHT BE ON PROPERTY. This legal document is which of the following?
    a. fire prevention code
    b. building code
    c. minimal standards
    d. elements and standards

57. As the company officer, you will use many leadership styles to accomplish the mission of the department and the goals of the company. You have an employee who is a five-year veteran and has demonstrated strength in skills and abilities on the job. This employee functions at the next level consistently and routinely takes on projects and requires little support from you. Which leadership style will you most likely utilize?
    a. directing
    b. consulting
    c. supporting
    d. delegating

58. The deliberate and apparent process by which one focuses attention on the communications of another is known as ______________.
    a. listening
    b. active listening
    c. passive listening
    d. aggressive listening

59. Incident priorities include which of the following?
    a. life safety
    b. incident stabilization
    c. property conservation
    d. all of the above

60. Company officers have status and power. We add this concept to our list of motivators that get employees to do what is asked of them. You ask a firefighter to take out the trash. The task is quickly completed. One reason would be that you, as the company officer, have ______ power.
    a. legitimate
    b. reward
    c. punishment
    d. identification

61. The leading cause of fire-related fatalities in the residential environment is _______________.

    a. arson
    b. smoking
    c. weather
    d. heating appliances

62. A disagreement, quarrel, or struggle between two individuals or groups is known as _______________.

    a. complaint
    b. grievance
    c. gripe
    d. conflict

63. As a company officer, you have assigned personnel. These personnel are dependent upon you for guidance daily. This makes your primary role _______________.

    a. leadership
    b. coaching
    c. mentoring
    d. friendship

64. You are a newly promoted lieutenant in your department. Where do you fit into the department's organization?

    a. middle management
    b. lower management
    c. first-line supervisor
    d. a & c
    e. b & c

65. You respond to an automatic alarm and are the second apparatus to arrive. Your standard assignment is to take the water supply and the Fire Department Connection. You do this based on the procedures that are written known as which of the following?

    a. quick access pre-plans
    b. notes
    c. standing operating guidelines
    d. none of the above

66. You are a newly promoted company officer. From where will most of your management occur?

    a. behind a desk
    b. in the field
    c. in the fire station
    d. none of the above

67. Collecting similar items into a single account and presenting them in a general area collectively under this topic or one line in a budget document is _______________ budget.

    a. capital
    b. operating
    c. program
    d. line-item

68. A formal statement that defines a course or method of action is _______________.

    a. procedure
    b. policy
    c. plan
    d. program

69. You respond to an automatic alarm and are the second apparatus to arrive. Your standard assignment is to take the water supply and the fire department connection. You do this based on the procedures that are written known as which of the following?

    a. quick access pre-plans
    b. notes
    c. standing operating guidelines
    d. none of the above

70. The second stage of fire is known as _______________ phase.

    a. incipient
    b. free-burning
    c. backdraft
    d. smoldering

71. The span of control is important as a company officer is responsible for human resources. According to the Company Officer text book, the number of personnel that a supervisor can effectively manage is _____ to ______.

    a. 2, 4
    b. 4, 7
    c. 3, 5
    d. 5, 8

72. If you are preplanning an old cotton mill, you would anticipate the construction to be _______________.

    a. ordinary
    b. Type III
    c. heavy timber
    d. Type V
    e. none of the above

73. Which of the following are proven time eaters?
    a. lack of personal goals and objectives
    b. good planning
    c. procrastination
    d. a & c
    e. b & c

74. As an applicant with a department, your employment is conditional on your joining the union. Failure to join or agreement to join resulting in your employment is an example of ______________ shop.
    a. open
    b. closed
    c. union
    d. agency

75. Communications that are more simple and spontaneous are known as _______________.
    a. formal
    b. informal
    c. written
    d. none of the above

76. During emergency scene operations, the firefighter typically would be the _______ in the communications model.
    a. sender
    b. message
    c. receiver
    d. all of the above

77. From 2002, the leading cause of death to firefighters was _______________.
    a. responding/returning from alarms
    b. training
    c. fire ground
    d. non-fire emergencies

78. Big words, technical terms, language differences are examples of _______________.
    a. mediums
    b. barriers
    c. understanding
    d. problems

79. You are a company officer in a department that has multiple stations. Your station is a multiple-company station and you are a newly promoted company officer on a rescue company. Your battalion chief asks you to develop a work plan for the next year. What type of plan would you be developing?
    a. long-range term
    b. mid-range plan
    c. short- range plan
    d. mini plan

80. The various maneuvers that can be used to achieve a strategy while fighting a fire or dealing with a similar emergency is known as _______________.
    a. tactics
    b. strategy
    c. size-up
    d. action plan

81. A document that attests that one has demonstrated the knowledge and skills necessary to function in a particular craft or trade is known as _______________.
    a. qualification
    b. training
    c. certification
    d. all of the above

82. You are the senior company officer and the first arriving company officer to the scene of a multi-family residential apartment complex on Halloween night. You arrive to find this two-story apartment complex with heavy fire involvement on the first and second floors of one end involving at least four apartments. Dispatch had a subject on the phone begging for help when the phone was dropped and the line remained open for a period of time before going dead. You have a victim unaccounted for at this time. You are making an interior attack. This would be known as _______________ mode.
    a. offensive
    b. defensive
    c. transitional
    d. none of the above

83. You are a company officer in a department that has multiple stations. Your station is a multiple-company station and you are a seasoned company officer on a rescue company. Your battalion chief asks you to develop a five-year station work plan. One year into this plan, you find that you are behind by two months. During the next year, you make up this deficit and are on schedule as planned. This is an example of _______________.
    a. controlling
    b. commanding
    c. organizing
    d. developing

84. Which of the following are not generally accepted tools for dealing with critical incident stress?
    a. training
    b. debriefing
    c. alcohol
    d. counseling
    e. scene management

85. As a fire continues to build, the smoke levels drop. When hot, fuel-rich gases meet fresh air, they ignite and burn along the ceiling. This is known as ______________.
    a. backdraft
    b. flashover
    c. Phase II
    d. rollover

86. You arrive on the fire scene and give a size-up. Upon entering the structure, you are advising command you are entering with four personnel. When you begin your attack, you announce water on the fire. These are examples of ______________.
    a. tactics
    b. strategies
    c. flow procedures
    d. benchmarks

87. An organizational directive that establishes a standard course of action is called a ______________.
    a. standard operating guideline
    b. standing operating procedure
    c. both a & b
    d. none of the above

88. Which of the following is the leading cause of injuries in fires?
    a. matches
    b. heating equipment
    c. cooking
    d. natural causes

89. The term company is used to describe work teams. The individual of that work team who must coordinate activities is the company officer. This individual has many roles to fill. Which of the following is not one of those roles?
    a. help in advancing hose
    b. directing crew operations
    c. activities that don't compromise roles as leaders
    d. developing a mission statement

90. An obstacle in communications that prevents the message from being understood by the receiver is a ______________.

    a. barrier
    b. reading
    c. active listening
    d. none of the above

91. The actions taken to prevent a fire from occurring or if one occurs to prevent a loss is known as?

    a. fire clean up
    b. fire attack
    c. fire prevention
    d. none of the above

92. The company officer is the crew leader and first-line supervisor of assigned staffing and equipment. We refer to these assigned personnel as ______________.

    a. staff
    b. bodies
    c. human resources
    d. facility resources

93. A team or company of emergency personnel kept immediately available for the potential rescue of other emergency responders is known as the ______________.

    a. safety team
    b. rapid response team
    c. rapid intervention team
    d. rapid safety team

94. A formal document indicating the focus, direction and values of an organization is ______________.

    a. vision
    b. mission statement
    c. SOPs
    d. rules and regulations

95. As a company officer, you are always looking for the advantage over emergency responses especially fire. Which one of the following would be classified as one of our greatest allies in fire protection?

    a. construction features
    b. fire codes
    c. sprinkler systems
    d. smoke detectors

96. A graphic representation of what the organization should look like and defines the formal lines of authority and responsibility is known as _______________.
    a. organizational chart
    b. organizational growth
    c. informal organization
    d. group chart

97. An administrative process whereby an employee is punished for not conforming to the organizational rules and regulations is known as _______________.
    a. acceptable
    b. commendation
    c. disciplinary action
    d. all of the above

98. As a company officer, in which of the following three essential elements in fire prevention would you be the least involved?
    a. education
    b. engineering
    c. enforcement

99. This is a target or other objective by which achievement can be measured. This also helps define purpose and mission.
    a. goal
    b. objective
    c. vision
    d. value

100. As the company officer, you will use many leadership styles to accomplish the mission of the department and the goals of the company. You have an employee who is a five-year veteran and has demonstrated progressing strength in skills and abilities on the job. Which leadership style will you most likely utilize?
    a. directing
    b. consulting
    c. supporting
    d. delegating

101. Your preplans program is underway and assignments have been made. You are instructed to use a plot plan for your drawing on the pre-incident survey sheet. Which of the following best describes the drawing portion?
    a. bird's-eye view of the property
    b. floor plan
    c. engineering schematic
    d. three-dimensional plan

102. You are the company officer determining an employee's competency and commitment. This employee has strong skills and self-confidence. They are always looking for ways to get involved or projects on which to work. This employee would be which of the following?
    a. high competency / high commitment
    b. low competency / high commitment
    c. high competency / low commitment
    d. low competency / low commitment

103. The ______ report should paint a concise, but vivid oral picture of the conditions you observe as well as a quick summary of your intentions and needs.
    a. brief
    b. oral
    c. written
    d. initial

104. Under the needs theory, a firefighter who you supervise takes responsibility for their efforts and has far exceeded the training requirements for two positions above their position. They have set even higher goals with you to continue to grow. You would say this person is which of the following?
    a. high-achievement needs
    b. high-affiliation needs
    c. low-achievement needs
    d. low-affiliation needs

105. The first step in the formal disciplinary process is a(n) ______________.
    a. disciplinary action
    b. written reprimand
    c. oral reprimand
    d. suspension

106. You are the company officer and an employee comes to you with a problem. They state they have a problem, but also have a potential solution. To assure you have received the complete message, you should practice which of the following?
    a. communications
    b. barrier breakdown
    c. active listening
    d. passive listening

107. Orders given on the emergency scene during operations by the company officer are in the form of ______________.
    a. informal
    b. formal
    c. oral
    d. a & c
    e. b & c

108. Place in order the following pieces of the command sequence: 1. size-up 2. implementing an action plan 3. developing an action plan
    a. 1, 3, 2
    b. 3, 2, 1
    c. 2, 3, 1
    d. 1,2, 3
109. You are inside a working structure fire and you encounter heat that is significant in nature near the 3-foot mark off of the floor. Below that mark, it is significantly cooler. In fire behavior, you know that the heat is an indication of the potentials that exist. We know this concept on higher temperatures closer to the ceiling as _______________.
    a. thermal stratification
    b. line of demarcation
    c. thermal plane
    d. thermal balance
110. As a company officer, in which of the following three essential elements in fire prevention would you be the most involved?
    a. education
    b. engineering
    c. enforcement
111. You respond to a reported structure fire in a university dormitory where there are known invalids. The plan for them is to protect in place or go to a portion of the structure that is relatively safe from fire and the by-products of fire. This area is known as an area of _______________.
    a. security
    b. safety
    c. congregating
    d. refuge
112. A traditional approach to organizing a company may be characterized where the company officer is reported to by all of his crew. This is known as which type of structure?
    a. linear organizational
    b. straight organizational
    c. flat organizational
    d. round organizational
113. As a company officer, you have the ability to get many tasks accomplished. When promoted, you were automatically given power through the position. By this power, you make employees work to levels to which they may not normally be self-motivated to work. This concept is known as _______________ power.
    a. reward
    b. legitimate
    c. punishment
    d. persuasion

114. You are a company office who arrives on the scene of a strip mall complex. You see heavy volumes of fire coming through the roof of an anchor store. As which type of building construction would this strip mall be classified?
    a. fire-resistive
    b. non-combustible
    c. heavy timber
    d. frame

115. A step in the disciplinary process that provides the employee a fresh start in another venue results in a _______________.
    a. transfer
    b. dismissal
    c. commendation
    d. empowerment

116. The organizational concept that refers to the uninterrupted series of steps or layers within an organization refers to which of the following?
    a. linear principle
    b. unity principle
    c. scaler principle
    d. division of labor principle

117. Place in order the following pieces of the command sequence: 1. size-up 2. implementing an action plan 3. developing an action plan
    a. 1, 3, 2
    b. 3, 2, 1
    c. 2, 3, 1
    d. 1,2, 3

118. The ABC fire department has a pyramid organizational structure. Where does the company officer fall within this structure?
    a. front-line supervisor
    b. first-line supervisor
    c. second-level supervisor
    d. a & b
    e. a & c

119. You and your company are the dispatched unit that command has designated as the rapid intervention team. Which of the following best describes your duties per the textbook?
    a. You are the team of firefighters which will have a minimum of two personnel. You are assigned the primary task of rescuing firefighters should the need arise. You will also be immediately available for any task that command has to assign to you.
    b. You are the team of firefighters which will have a minimum of two personnel. You are assigned the primary task of rescuing firefighters should the need arise. You will not be immediately available for any task that command assigns to you.

120. The difference between the fire triangle and the fire tetrahedron is _______________.
   a. heat
   b. fuel
   c. chemical chain reaction
   d. oxygen

121. Certification means that an individual has been tested by an accredited examining body on clearly identified materials and found to meet the minimal standard. There are other reasons to certify. Which of the following is not one of the reasons to certify?
   a. protection from liability
   b. recognition of demonstrated proficiency
   c. recognition of professionalism
   d. all of the above
   e. b & c only

122. America Burning was published in _______________.
   a. 1903
   b. 1943
   c. 1973
   d. 1993

123. The strategically combined various maneuvers that are used to achieve a strategy while fighting a fire or dealing with a similar emergency is known as _______________.
   a. tactics
   b. strategy
   c. size-up
   d. action plan

124. Activities that take about as long as the time is allowed is known as a concept called the _______________.
   a. Managerial Law
   b. Parkinson's Law
   c. Range Law
   d. Empowering Law

125. William Ouchi developed the management style in which the manager believes that people not only like to work and can be trusted, but that they want to be collectively involved in the management process and recognized when successful. This is known as which of the following theories?
   a. Theory A
   b. Theory X
   c. Theory Z
   d. Theory Y

126. The National Fire Protection Standard for Fire Officer is ______________.
    a. NFPA 1021
    b. NFPA 1201
    c. NFPA 1001
    d. NFPA 1403

127. Professionalism encompasses attitude, behavior, communication, style, demeanor, and ethical beliefs. Which of the following best describes attitude?
    a. the core of your performance
    b. nothing that helps job performance
    c. suggests you are working to be a good role model
    d. all of the above
    e. a & c

128. You have been a company officer for five years and have been found to be reliable and dependable to the higher administrative staff on assigned projects. The more you have proven your talents, the more special assignments you seem to be getting pushed your way. Now you are finding you are struggling to get routine tasks accomplished. Which one of the following best describes you in the proven time eaters?
    a. lack of personal goals
    b. reacting to urgent events
    c. inability to say no
    d. trying to do too much yourself

129. As a company officer, you are faced with a question for which you are not sure of the answer. What should you do?
    a. Turn it back to the employee to find the answer.
    b. Admit you don't know the answer.
    c. Try to find the answer.
    d. a & b
    e. b & c

130. During emergency scene operations, the firefighter typically would be the _______ in the communications model.
    a. sender
    b. message
    c. receiver
    d. all of the above

131. A financial plan to acquire the goods and services needed to run an organization for a specific period of time is referred to as the ______________ budget.
    a. line-item
    b. program
    c. capital
    d. operating

132. As a company officer, upon promotion you are afforded the right and power to command. This right and power by position is known as _______________.
   a. respect
   b. power
   c. managerial legal right
   d. authority

133. The definition of ________ is being responsible for one's personal activities; in the organizational context, it includes being responsible for the actions of one's subordinates.
   a. responsibility
   b. accountability
   c. line authority
   d. management

134. Fire officers must base their actions on priorities such as life safety, property conservation, and incident stabilization. Which of the following puts these in the correct order?
   a. Life safety, incident stabilization, property conservation
   b. Incident stabilization, life safety, property conservation
   c. property conservation, life safety , incident stabilization
   d. property conservation, incident stabilization, life safety

135. Company officers are expected to be a _______ to their assigned personnel.
   a. full-time instructor
   b. full-time leader
   c. both a & c
   d. none of the above

136. A financial plan to purchase high-dollar items that have a life expectancy of more than one year is referred to as the _______________ budget.
   a. operating
   b. line-item
   c. capital
   d. program

137. You are the shift training officer and you have been given a house in which to do live fire training. You will be following NFPA 1403. Your structure is a single-story house with 2,210 square feet. It has a large fireplace in the living room and a basement with narrow steps leading down off of the kitchen. What will be required to make this structure safe?
   a. floors, railing and stairs made safe
   b. asbestos removed
   c. chimney hazards eliminated
   d. extraordinary dead weight removed
   e. all of the above

138. You have an employee who you feel doesn't like work and you have found that he cannot be trusted to keep deadlines or meet minimum productivity levels. You are describing which of the following management theories?
    a. TQM
    b. Theory X
    c. Theory Y
    d. Theory Z

139. The organizational or command principle whereby there is only one boss no matter how many different organizations, divisions or groups is referred to as ______________.
    a. unity of command
    b. command
    c. control
    d. none of the above

140. The command sequence consists of which of he following? 1. life safety 2. developing an action plan 3. implementing the action plan 4. pre-incident planning 5. post-incident analysis
    a. 1, 3, 4
    b. 1, 2, 4
    c. 1, 2, 3
    d. all of the above

141. A characteristic or organizational structure denoting the relationship between supervisors and subordinates is called ______________.
    a. accountability
    b. line authority
    c. responsibility
    d. flow chart

142. You respond to a residential structure fire that has been vacant and for rent over the last year. The electrical meter base is capped with no meter in place. The fire started on the second floor. The rear door is open and you find strange patterns of char on the floor with close areas not burned. What is you assumption on the cause of this fire?
    a. electrical cord under the rugs
    b. arson
    c. electrical wiring fire in the floor
    d. none of the above

143. Lightweight scissor wood trusses are commonly used in which of the following types of construction?
    a. Type I
    b. Type II
    c. Type III
    d. Type IV
    e. Type V

144. Solutions to the nation's fire problem rest with ongoing fire prevention activities. Which of the following is not an essential element of fire prevention?
    a. education
    b. engineering
    c. enforcement
    d. environment

145. The final step in the management process involves monitoring the process to ensure that the work is accomplishing the intended goals and objectives and taking corrective action when it is not. This is known as ______________.
    a. coordinating
    b. organizing
    c. controlling
    d. commanding

146. Which of the following is not one of the four career development roles of a leader?
    a. coach
    b. apprentice
    c. adviser
    d. referral agent

147. An opportunity to take a short break from firefighting duties to rest, cool off, and rehydrate is known as ______________.
    a. rehabilitation
    b. reconciliation
    c. rest period
    d. none of the above

148. Company officers should remember that they have a right to complete an investigation, and that they should have continuous presence on the property until that investigation is complete or an investigative team has released them. Once the premises is vacated, what has to be secured prior to reentering the premises to do an investigation?
    a. probable cause
    b. search warrant
    c. nothing has to occur
    d. administrative process

149. Communications that usually have legal standing within the organization are known as ______________.
    a. informal
    b. written
    c. informational
    d. formal

150. The NFPA standard for Fire Fighter Professional Qualifications is ______________.
   a. 1001
   b. 1002
   c. 1021
   d. 1710

151. As a company officer, you meet with your crew and set a vision for where the company is going based on the department's mission statement. You set clear concise expectations based upon the guiding principles of the department. In developing this, you are focusing on both the internal and external customer's needs. In doing this, you are practicing what management style?
   a. Theory X
   b. Theory Z
   c. traditional management
   d. total quality management

152. The NFPA standard for live fire training is ______________.
   a. 1001
   b. 1002
   c. 1017
   d. 1403

153. A mental assessment of the situation; gathering and analyzing information that is crucial to the outcome of the event is ______________.
   a. pre-planning
   b. pre-incident analysis
   c. size-up
   d. none of the above

154. Use the Iowa State formula. You respond to a commercial structure which is 150 feet wide, 500 feet long, and 12 feet high on each of 50 floors. With this knowledge figure fire flow based on an appropriate theoretical fire flow formula utilizing all of the building's dimensions. How much would the fire flow be (in gpms) if the building was 25% involved on the 50th floor?
   a. 10,000 gpm
   b. 2,250 gpm
   c. 3,000 gpm
   d. 30,000 gpm

155. The management theory in which the focus of the organization is on the continuous improvement geared to customer satisfaction is known as ______________.
   a. total quality management
   b. Theory A
   c. Theory Z
   d. Theory X

156. A system of values or a standard of conduct is referred to as _______________.

  a. laws
  b. standards
  c. ethics
  d. principle

157. You are a company officer who is in charge of a crew stationed in a mixed district representing some business and residential response potential. In the response district, the residential configurations range from million dollar single family homes to multi-family housing to mobile homes. This mixture will bring a diverse group of residents that depend on you for services. To understand all of the population, it is important to have a crew strong in which of the following areas?

  a. emergency scene operations
  b. cultural diversity
  c. tactical considerations
  d. all of the above

158. Self-operating thermosensitive device that releases a spray of water over a designed area to control or extinguish a fire is known as _______________.

  a. deluge gun
  b. stand pipe
  c. pre-piped system
  d. automatic fire protection sprinkler

159. As a company officer, you will receive a larger work load and responsibility with the position. Often times, tasks are assigned to the company officer to complete. As work loads increase, it is often easy to become overloaded. This is due to the officer not delegating appropriate assignments and trying to do the entire assignment themselves. This occurs when officers will say "I can do it better myself," "It will take too much time to train someone to do that task," or "I really can't take a chance on that project." These are examples of _______________ delegation.

  a. barriers to
  b. reasons for
  c. hemisphere of
  d. facts about

160. Which of the following is not a role of the company officer?

  a. coach
  b. counselor
  c. role model
  d. administrator
  e. none of the above

161. To give authority or power to another is _______________.
   a. empowerment
   b. powerment
   c. leadership
   d. termination

162. You have been a company officer for five years and have been found to be reliable and dependable to the higher administrative staff on smaller assigned projects. The Chief has asked you to head up a hiring process for new recruits. It is summer time and you have a new boat you and your family are enjoying everyday you are off. You also have taken several days off to go to the lake. You were assigned the hiring project on June 9. The completion date has been set for July 23. Today, it is July 13 and you have done very little work on this project. Which one of the following best describes you in the proven time eaters?
   a. lack of personal goals
   b. reacting to urgent events
   c. procrastination
   d. trying to do too much yourself

163. The critical process of shifting from the offensive mode to the defensive mode or the defensive mode to the offensive mode is known as _______________ mode.
   a. switching
   b. moving
   c. transitional
   d. flip flop

164. You are the senior company officer and the first arriving company officer to the scene of a multi-family residential apartment complex on Halloween night. You arrive to find this two-story apartment complex with heavy fire involvement on the first and second floors of one end involving at least four apartments. Dispatched had a subject on the phone begging for help when the phone was dropped and the line remained open for a period of time before going dead. You now have brought the fire under control. You have a victim unaccounted for at this time Which of the following meet the guidelines for calling a fire investigator?
   a. potentially incendiary fire
   b. a fire resulting in a death
   c. property damage greater than $50,000
   d. a & b
   e. all of the above

165. Which of the following is not one of the three major components of customer service?
   a. Constantly be selective of the services.
   b. Always be nice.
   c. Regard everyone as a customer.
   d. Constantly raise the bar.

166. The act of assigning duties to subordinates is ______________.

   a. command
   b. entrepreneurship
   c. delegation
   d. dumping

167. ABC fire department has a progression or career development path for individuals to move up through the ranks from firefighter to chief. This uninterrupted series of steps or layers is an organizational concept known as ______________.

   a. flat principle
   b. scaler principle
   c. unity of command
   d. division of labor

168. You are reviewing an employee's performance evaluation for the quarter. It is important to have good communications occurring. This two-way communications allows the employee to provide ___ on the evaluation.

   a. feedback
   b. knowledge
   c. barriers
   d. all of the above

169. The communications model has several steps. The process ends with ______________.

   a. sender
   b. receiver
   c. feedback
   d. medium

170. A dramatic event in a room fire that rapidly lead to full involvement of all combustible materials present almost simultaneously is known as ______________.

   a. backdraft
   b. rollover
   c. flashpoint
   d. flashover

171. An assessment of the safety hazards for both civilians and firefighters in a particular occupancy lightweight truss construction is known as the ______________ factors.

   a. survival
   b. enhancement
   c. structural
   d. occupancy

172. As a newly promoted company officer, you are _______________.
   a. responsible for the performance of the assigned crew
   b. responsible for the safety of the assigned crew
   c. second-line supervisor
   d. a & b
   e. all of the above

173. The act of controlling the fire to prevent the fire from extending any further in the structure is known as _______________.
   a. fire detection
   b. overhaul
   c. extinguishment
   d. confinement

174. The accomplishment of the organization's goals by utilizing the resources available is done through the leadership of _______________.
   a. management
   b. command
   c. unity of direction
   d. Fayol

175. A portion of a structure that is relatively safe from fire and the products of combustion typically used for protecting occupants in place is known as _______________.
   a. floor area
   b. safe haven
   c. holding area
   d. area of refuge

176. The on-going process of evaluating the incident throughout its duration is known as _______________.
   a. strategizing
   b. tactic deployment
   c. action planning
   d. size-up

177. The addition of a sprinkler system to a building lowers which of the following?
   a. life risk factors
   b. property risk factors
   c. physical risk factors
   d. community risk factors

178. As a company officer, it is necessary to provide a good work environment for your employees. Herzberg's model describes what is needed to get employees to come to work and not be dissatisfied. Items like good working relationships and considerate supervisors are examples of _______________.

a. hygiene factors

b. motivators

c. achievements

d. power

179. You have an employee who brings in a pornographic magazine to work. He shows a picture of a female in the magazine to several members of the crew. He states the female in the picture looks like another department employee. What answer best describes the condition described here?

a. harassment

b. sexual harassment

c. Civil Rights Act violation

d. b & c

e. all of the above

180. An organizational principle that addresses the number of personnel a supervisor can effectively manage is the _______________.

a. command

b. unity of command

c. delegation

d. span of control

181. You as the company officer have been instructed by the fire chief to write a letter to a citizen referencing a complaint about the use of sirens during the middle of the night. Which of the following is important?

a. technical terms

b. emphasis

c. considering the reader

d. complexity

182. Which of the following is not a leadership style?

a. directing

b. consulting

c. supporting

d. delegating

e. advocating

183. You are assigned as the leader of a task force that will be working in a collapsed building. You have selected team members that are specifically trained in the area of structural collapse with specialties in shoring, engineering design, and fiber optic visual technology. As the task force leader, you would be consider to be___________ this group as you begin rescue operations for trapped victims.

 a. organizing
 b. commanding
 c. developing
 d. coordinating

184. The company officer is responsible for the _________ of the assigned personnel in an emergency services organization

 a. performance
 b. safety
 c. mission development
 d. both a & b
 e. both a & c

185. The principle job of the company officer is to _______________.

 a. write reports
 b. lead others
 c. fight fire
 d. emergency response

186. You have a large retirement facility with skilled nursing facility included. This type of complex is an example of a _______________.

 a. fire target
 b. target hazard
 c. facility with multiple automatic alarms daily
 d. none of the above

187. You are senior company officer who has been tasked at looking into the health and safety program of the department. You are tasked with bringing the department's rules, regulations, and standard operating guidelines into compliance with professional standards and laws. Which two organizations would you look to for information and guidance?

 a. JEMS and FIREHOUSE Magazines
 b. NIOSH and JOCCA
 c. NFPA and JOCCA
 d. NFPA and OSHA

188. Firefighter deaths by nature was topped by _____ in 2002?

 a. asphyxiation
 b. heart attacks
 c. burns
 d. internal trauma

189. Which NFPA standard states that fire departments should have a physical fitness program?

 a. 1001

 b. 1002

 c. 1021

 d. 1500

190. Which of the following is the first step in the management process involving looking to the future and determining objectives?

 a. organizing

 b. planning

 c. commanding

 d. coordinating

191. You are the investigating officer on a small room and contents fire in a bedroom. The fire began about 11 PM on Friday, February 1. The temperature outside is -10 degrees and there is 8 inches of snow on the ground. You have conducted interviews with the occupants who are smokers and are now looking at findings in the room. The bed is burned significantly and the heat line of demarcation is about 5 feet off the floor. There is heavy smoke residue in the room with minor fire damage to any other furniture. You find a space heater plugged in up close to the window beside the bed, several extension cords, and an ashtray by the bed. What would you suspect the cause of this fire to be?

 a. electrical

 b. improperly positioned space heater

 c. smoking

 d. not enough information to determine

192. The typical downtown business districts in older communities where the buildings are generally no more than five stories tall and the exterior is masonry is generally known as which type of building construction?

 a. Type I

 b. Type II

 c. Type III

 d. Type IV

193. You are on the second alarm responding to a large building fire. Knowing you will be the first engine arriving on the scene in the second alarm, the incident commander summons you to the command post and assigns your crew to work as part of a task force with another company. You are given the task of tracking personnel and their location during the incident. This process is known as _______________.

 a. incident command

 b. unified command

 c. personnel accountability

 d. personnel management

194. As a company officer, you must look at the needs for your crew and develop them accordingly. The first step in this management process would be _______________.

a. planning

b. delegating

c. centralizing

d. advocating

195. A company officer certified as a Fire Officer I should meet the National Fire Protection Association Standard 1021 ________.

a. Chapter 2

b. Chapter 3

c. Chapter 4

d. Chapter 5

196. As a company officer, you will set levels of achievement with your employees that can be effectively measured to assure that achievement has occurred. In setting these levels, you are helping the employee set obtainable _______________.

a. objectives

b. JPRs

c. progress

d. goals

197. You are the senior company officer and the first arriving company officer to the scene of a multi-family residential apartment complex on Halloween night. You arrive to find this two-story apartment complex with heavy fire involvement on the first and second floors of one end involving at least four apartments. Dispatch had a subject on the phone begging for help when the phone was dropped and the line remained open for a period of time before going dead. You have a victim unaccounted for at this time. On arrival, what are your sequential incident priorities based upon response to this scenario?

a. life safety, incident stabilization, property conservation

b. incident stabilization, property conservation, life safety

c. life safety, property conservation, incident stabilization

d. a or b

e. all of the above

198. A person assigned as manager of the department's health and safety program is known as the _______________.

a. company officer

b. chief

c. health and safety institute

d. health and safety officer

199. You are the company officer who arrives on scene of a 30-story building with fire showing from one window on the 22nd floor. The building described here will most likely be made of which type of construction?

 a. fire-resistive
 b. non-combustible
 c. woodframe
 d. ordinary

200. You are a company officer in a department that has multiple stations. Your station is a multiple-company station and you are a seasoned company officer on a rescue company. Your battalion chief asks you to develop a five-year station work plan. What type of plan would you be developing?

 a. long-range term
 b. mid-range plan
 c. short- range plan
 d. mini plan

Final Exam: Answers to Questions

1. T
2. T
3. T
4. F
5. T
6. T
7. T
8. T
9. T
10. T
11. F
12. T
13. T
14. T
15. F
16. F
17. T
18. F
19. T
20. F
21. C
22. C
23. C
24. B
25. A
26. B
27. D
28. B
29. D
30. C
31. D
32. E
33. A
34. A
35. C
36. C
37. A
38. A
39. C
40. C
41. B
42. A
43. D
44. E
45. D
46. B
47. D
48. B
49. C
50. D
51. D
52. D
53. B
54. D
55. A
56. A
57. D
58. B
59. D
60. C
61. B
62. D
63. A
64. E
65. C
66. B
67. D
68. B
69. C
70. B
71. B
72. C
73. D
74. B
75. B
76. C
77. C
78. B
79. C
80. A
81. C
82. A
83. A
84. C
85. D
86. D
87. B
88. C
89. D
90. A
91. C
92. C
93. C
94. A
95. C
96. A
97. C
98. B
99. A
100. B
101. A
102. A
103. D
104. A
105. C
106. C
107. E
108. A
109. A
110. A
111. D
112. C
113. B
114. B
115. A
116. C
117. A
118. D
119. B
120. C
121. D
122. C
123. D
124. B
125. C
126. A
127. A
128. C
129. E
130. C
131. D
132. D

133. B
134. A
135. C
136. C
137. E
138. B
139. A
140. C
141. B
142. B
143. E
144. D
145. C
146. B
147. A
148. B
149. D
150. A
151. D
152. D
153. C
154. B
155. A
156. C
157. B
158. D
159. A
160. D
161. A
162. C
163. C
164. E
165. A
166. C
167. B
168. A
169. C
170. D
171. A
172. D
173. D
174. A
175. D
176. D
177. A
178. A
179. E
180. D
181. C
182. E
183. B
184. D
185. B
186. B
187. D
188. B
189. D
190. B
191. B
192. C
193. C
194. A
195. C
196. D
197. A
198. D
199. A
200. B

Final Exam: Rationale & References for Questions

Question #1. Each worker's compensation claim is money paid out. NFPA 5.4.2, 5.7.1. *CO, 2E,* Page 177.

Question #2. Effective communications are a vital part of your professional life. NFPA 4.1.2. *CO, 2E,* Page 20.

Question #3. This is a cycle due to the achievements and progression and growth of personnel. NFPA 5.2.2. *CO, 2E,* Page 152.

Question #4. The fire service is typically a team organization in every aspect. NFPA 4.2.2, 5.2. *CO, 2E,* Page 125.

Question #5. This is due to the fact that most firefighters die early on in incidents and hose advancement is generally in the early stages of incidents. NFPA 4.7.2, 5.7.1. *CO, 2E,* Page 179.

Question #6. It is important to be able to manage time to have good efficiency. NFPA 4.7 5.7.1. *CO, 2E,* Page 105.

Question #7. Most officers enjoy the challenges with which they are faced because, to get to the officer level, you must be a high achiever. NFPA 5.2 4.2.2. *CO, 2E,* Page 125.

Question #8. Most officers have high-power needs. NFPA 5.2. *CO, 2E,* Page 125.

Question #9. The fire officer helps meet all of the objectives and the grading of ISO. NFPA 5.1.1. *CO, 2E,* Page 152.

Question #10. Most fires are not investigated by an investigator, but the company officer. NFPA 4.5.1, 4.5.2, 5.5.2. *CO, 2E,* Page 276.

Question #11. Lightweight construction is in Type II or Type V based on if it is wood or steel. NFPA 4.6.1. *CO, 2E,* Page 251.

Question #12. The officer helps the firefighter develop based on the community needs. NFPA 4.1. *CO, 2E,* Page 12.

Question #13. Arson has to be targeted at someone. NFPA 4.5.1, 4.5.2, 5.5.2. *CO, 2E,* Page 271.

Question #14. Most of the time, officers deal with other functions rather than emergency response. NFPA 4.2. *CO, 2E,* Page 5.

Question #15. Every person in the fire service should live by the fire service code of ethics. NFPA 4.3.2, 4.3.3. *CO, 2E,* Page 91.

Question #16. Being able to write is an important part of your professional career. NFPA 4.1.2. *CO, 2E,* Page 28.

Question #17. With change comes confusion and uneasiness. if officers can keep communication open and flowing, there is an excellent chance that change will be easier. NFPA 4.2.2, 4.2.3, 5.4.1. *CO, 2E,* Page 166.

Question #18. Planning is a systematic process and should be ongoing. NFPA 5.2.1, 4.2.6, 4.2.1. *CO, 2E,* Page 99.

Question #19. Working with others to accomplish thc mission of the department is the focus of the company officer. NFPA 5.2.1; 4.2.1. *CO, 2E,* Page 5.

Question #20. Most structured departments have a pyramid-style design in which it is important to be able to communicate from top to bottom and bottom to top. NFPA 4.2. *CO, 2E,* Page 60.

Question #21. Target hazards are those occupancies that present high risk to life safety and property. Given that you cannot plan for every situation, you should plan for those that present the greatest risk. NFPA 4.6.2 5.6.1. *CO, 2E,* Page 293.

Question #22. The inability to say no or decline the opportunity is the hardest of the time eaters for most. You must remember to tie yourself to goals and focus on priorities. It is important to know when you are overloading your plate. Procrastination is just the opposite. You do not have a focus on goals or priorities as it relates to work. The continued putting off of this project has you in a position of crunch time for completion. This usually leads to poor work. NFPA 5.2.1. *CO, 2E,* Page 106.

Question #23. This is one of several leadership tools that focuses the employee's toward improving their work performance. NFPA 4.2.5. *CO, 2E,* Page 149.

Question #24. An important part of fire prevention and the company officer's job is to enforce codes that are applicable to the jurisdiction. Many of these cross between building and fire prevention codes. It is important to understand how each relates. NFPA 5.5.1. *CO, 2E,* Page 234.

Question #25. The importance of written communications is imperative. Many of these documents will be official records and will also serve as information being sent to the public. NFPA 4.3.3; 5.2.2; 5.2.1. *CO, 2E,* Page 28.

Question #26. Building construction will define fire behavior and the tactics required to extinguish a fire. NFPA 4.6.2. *CO, 2E,* Page 248-251.

Question #27. It is important to understand the infrastructure of the community and government of which the fire department is a part. NFPA 4.3.1. *CO, 2E,* Page 54.

Question #28. A plot plan will help you identify access points, exposures, and potential strategic set up locations. NFPA 4.6.1. *CO, 2E,* Page 295.

Question #29. This style of construction is referred to as Main Street, USA. NFPA 4.6.2. *CO, 2E,* Page 250.

Question #30. A good leader must understand human behavior and the needs of employees as the life outside of the workplace often influences life in the work atmosphere. NFPA 4.7; 5.7. *CO, 2E,* Page 123.

Question #31. The amount of water that is required theoretically to suppress the fire can be figured by both of these formulas. NFPA 4.6.1. *CO, 2E,* Page 263-264.

Question #32. Safety comes from the elimination of unsafe equipment and procedures, having and enforcing policies that provide for safety, and having viable recognition programs at all levels that reinforce both the company's and industry's commitment to worker safety and health. NFPA 4.7.1; 4.7.2; 5.7.1. *CO, 2E,* Page 178-179.

Question #33. As an incident commander, it is important to understand the work progress made during an incident. Benchmarking helps the incident commander keep track of tasks completed and the location of crew members for personnel accountability. NFPA 4.6, 4.7 5.6 5.7. *CO, 2E,* Page 320.

Question #34. Groups exist whenever two or more people share a common goal. Organizations are groups of people. Typically, these people share a common goal, have formal rules, and have designated leaders. This is true of fire companies. NFPA 4.2. *CO, 2E,* Page 44-46.

Question #35. Knowing the leading causes of fire as being careless smoking and the evidence presented, you would tend to lean toward that answer based upon fire behavior and the time of day, coupled with the ashtray finding and the smoking resident. NFPA 4.5.1. *CO, 2E,* Page 221.

Question #36. The accomplishment of the organization's goals by utilizing the resources available is management. As a company officer, you manage human resources daily. NFPA 4.2.2. *CO, 2E,* Page 75.

Question #37. The first stage of the fire is limited to the materials originally ignited. NFPA 4.6.2; 5.6.1. *CO, 2E,* Page 257.

Question #38. Fire suppression is a reaction: Resources are mobilized following an event and the action to mitigate an event. NFPA 4.6.3. *CO, 2E,* Page 214.

Question #39. The company officer must be an effective communicator whereas he is giving orders, receiving orders and dealing both with the public and the crew members. Without effective communications, the company officer can create impressions that are either false or obscured. NFPA 4.2. *CO, 2E,* Page 39.

Question #40. As a company officer, understanding fire growth is important as you make multiple split-second decisions on tactics and operations based upon the knowledge of fire behavior. NFPA 4.6.2. *CO, 2E,* Page 272.

Question #41. Coaching is an informational process that helps subordinates improve their skills and abilities. Coaching implies one-on-one relationship that treats subordinates as full partners. NFPA 5.2.1. *CO, 2E,* Page 147.

Question #42. The incident commander and the company officer are both responsible for these items whereas the initial arriving company officer is the incident commander for at least a few minutes. These priorities are how action plans are formulated. NFPA 5.6.1. *CO, 2E,* Page 320.

Question #43. Fayol is known as the father of management. He is known for his research and development of managerial principles. NFPA 5.2.1. *CO, 2E,* Page 77.

Question #44. The experience of a fire investigator will prove to be beneficial especially with the elements of potential death and potential arson. NFPA 4.5.1; 4.5.2; 5.5.2. *CO, 2E,* Page 277.

Question #45. Mediums can have a variety of problems including noise, language, terminology, and others. It is important to try to control this area as much as possible so your communications will be as effective and as clear as possible. NFPA 4.3.3; 4.2.1; 5.2.1; 5.2.2. *CO, 2E,* Pages 21.

Question #46. Length x Width x Height / 100 = the fire flow for each floor. If the floor is 1/4 involved, then divide that answer by 4 and you would get 1,500 gpm. NFPA 4.6.1. *CO, 2E,* Page 264.

Question #47. A goal is a target or object by which achievement can be measured. As company officers, we set departmental, company, and individual goals for ourselves and others daily. NFPA 4.2.3. *CO, 2E,* Page 98.

Question #48. Strategies are fundamental to planning and directing operations. NFPA 4.6, 5.6. *CO, 2E,* Page 333.

Question #49. Being a role model means being professional. It is important to have all of the necessary qualities to be an effective leader. NFPA 4.2; 5.2. *CO, 2E,* Page 15.

Question #50. This is a serious issue and usually occurs when the condition reaches the seriousness such that complaining doesn't seem to help fix or solve the issue or bring any relief. NFPA 5.2.2. *CO, 2E,* Page163.

Question #51. Although he fire service is a formal organization, informal relationships exist as well just as described in this question where a superior officer has an informal lunch with the crew of a station. NFPA 4.2.1; 4.2.2; 4.2.3. *CO, 2E,* Page 60.

Question #52. The company officer is responsible for many items. Operations relate for 10 percent and administrative aspects 90 percent whereas we do reports, evaluations, and other administrative document record keeping. NFPA 4.2.1, 4.2.2. *CO, 2E,* Page 5.

Question #53. The employer should encourage participation and growth along with provide the environment for this to occur. The individual must take advantage of the opportunities to enhance the skills and knowledge for advancement. NFPA 4.1.1. *CO, 2E,* Page 120.

Question #54. During an offensive mode, operations are conducted in an aggressive manor. NFPA 4.6, 5.6. *CO, 2E,* Page 327.

Question #55. Groups exist whenever two or more people share a common goal. Organizations are groups of people. Typically, these people share a common goal, have formal rules, and have designated leaders. This is true of fire companies. NFPA 4.2. *CO, 2E,* Page 44-46.

Question #56. This is the document by which you inspect. These codes are adopted by local, state, or other jurisdictional entities. NFPA 5.5.1. *CO, 2E,* Page 228.

Question #57. Most employees described in the question require little discussion, direction, and supervision. They are at the next level waiting for challenges and opportunities to present themselves. NFPA 5.2.1, 4.2.1. *CO, 2E,* Page 128.

Question #58. In active listening, the listener can actually show the sender that they are listening by focusing all of their attention on the speaker and showing a genuine interest in the message. Active listening is a significant portion of good communications. NFPA 5.2 ,4.2. *CO, 2E,* Page 24.

Question #59. Strategy and tactics are set to accomplish the goals for the incident. The priorities are set so as to provide a guide to what areas and tactics need to be taken care of in a sequential order if the situation dictates. NFPA 4.6, 5.6. *CO, 2E,* Page 320.

Question #60. The recognition of the power of position accompanied by the authority by virtue of the supervisor's ability to administer punishment leads others to see the punishment power concept. NFPA 5.2.1. *CO, 2E,* Page 127.

Question #61. Careless use and discarding of smoking materials is the leading cause of fire-related fatalities in the residential setting. NFPA 5.5. *CO, 2E,* Page 222.

Question #62. When individuals work together there will always be friction of dislike by someone for a variety of reasons. This behavior is a conflict. NFPA 5.2.1; 5.2.2. *CO, 2E,* Page 163.

Question #63. Leading others is the company officer's primary job as you are the one who is assigned to direct the personnel assigned to you. To accomplish the mission, you must lead others. NFPA 4.2.1. *CO, 2E,* Page 5.

Question #64. Company officers, for firefighters, are the first step in the chain of command. Command level officers consider the company officer to be the lower management staff and the first line supervisors for the organization. NFPA 4.2. *CO, 2E,* Page 4.

Question #65. This is a standardized way of functioning based upon an organized directive that establishes a standard course of action. NFPA 4.6.3; 5.6.1. *CO, 2E,* Page 304.

Question #66. As most fire chief's would agree that the manage primarily from behind a desk, they would also agree that the company officer is a working foreman managing personnel from where the action is occurring just like a floor supervisor in an industrial plant. NFPA 5.6.1; 4.6.3. *CO, 2E,* Page 84.

Question #67. Line-item budgeting is like keeping accounting on the money spent. This is used when you need specifics to be purchased and don't want it to be generic in nature. NFPA 4.4.3; 5.4.2. *CO, 2E,* Page 103.

Question #68. Management can be enhanced through good policy, procedures, and even by personal observation. Policies are broad in nature and should be clearly understood. NFPA 4.3.1; 5.2.1. *CO, 2E,* Page 84.

Question #69. This is a standardized way of functioning based upon an organized directive that cstablishes a standard course of action. NFPA 4.6.3; 5.6.1. *CO, 2E,* Page 304.

Question #70. Flames become visible along with an increased amount of heat and light generation. NFPA 4.6.2, 5.6.2. *CO, 2E,* Page 257.

Question #71. As a company officer, it is important to be able to manage incidents and daily functions. The use of span of control is important to be able to effectively manage these personnel. NFPA 4.2. *CO, 2E,* Page 67.

Question #72. These structure are called Mill structures or heavy timber based on the fact that the majority of these structures that were built, were mills. NFPA 4.6.2. *CO, 2E,* Page 250.

Question #73. People allow important events to become urgent events through the lack of good time management techniques. Putting off work creates rushes and inefficiency. Thus, this precludes lower quality in work performance and results in hurting the organization. NFPA 4.4.2; 5.2.1. *CO, 2E,* Page 106.

Question #74. As an organization that has a bargaining union, it is important for a company officer to understand the factors that surround the contract and the department so as not to violate any laws or conditions that are legally binding. It is important also to know these as your are the representative for your employees' best welfare. NFPA 5.2.1. *CO, 2E,* Page 102.

Question #75. We write with less formality when sending emails, memos, and short notes. We also tend to be less formal at social events. NFPA 5.2.1; 4.2.1; 4.2.2. *CO, 2E,* Page 20.

Question #76. As the company officer gives orders on the emergency scene, the firefighter, who is the receiver of the message, interprets the message and acts from there. NFPA 4.2.2. *CO, 2E,* Page 21.

Question #77. It is important to understand the causes of injuries and death of firefighters to take a proactive approach to reduce these numbers. NFPA 4.7.1; 4.7.2. *CO, 2E,* Page 179.

Question #78. There are many barriers to effective communications. These are some of the common ones. NFPA 5.2; 4.2. *CO, 2E,* Page 24.

Question #79. The first step in the management process is planning and the effectiveness of a company officer is the ability to make and keep plans moving forward. NFPA 5.4.5. *CO, 2E,* Page 77.

Question #80. These operations must be focused on a specific location and measurable as to their success. NFPA 4.6, 5.6. *CO, 2E,* Page 333.

Question #81. Certification means that an individual has been tested by an accredited examining body on clearly identified material and found to meet the minimum standard. NFPA 3.3.11. *CO, 2E,* Page 8.

Question #82. The priorities for any incident are as follows in order: life safety, incident stabilization, and property conservation. Operations that make a direct attack on the fire are in the offensive strategy to control the fire and keep it in place. NFPA 4.6.3. *CO, 2E,* Page 327.

Question #83. Controlling allows us to measure the effectiveness of our effort to help us maintain our goals. By doing so, we can seek ways to improve, thus increasing productivity. NFPA 5.6.1; 5.4.2. *CO, 2E,* Page 79.

Question #84. Critical incident stress is a significant part of an emergency responder's life. Assistance is often necessary. NFPA 4.6.4. *CO, 2E,* Page 206.

Question #85. It is important to understand the growth and progression of fire in developing action plans. NFPA 4.6.2. *CO, 2E,* Page 258.

Question #86. Benchmarking allows command to follow sequential progression in the incident and also note the priorities that are being met. NFPA 4.6.3. *CO, 2E,* Page 320.

Question #87. In certain cases, it is important to have a set plan of action to handle certain situations. SOPs allow for just that. NFPA 4.7, 5.7. *CO, 2E,* Page 304.

Question #88. Understanding the nation's fire problem will help you focus your goal setting toward helping make change. NFPA 4.5.1, 5.5.2. *CO, 2E,* Page 222.

Question #89. Company officers are expected to lend a hand when needed, whether it is advancing hose or other emergency scene operations so long as it doesn't compromise role as a leader. The company officer is not setting the mission or developing it. They are the ones working with the crews to meet or accomplish it. NFPA 4.2.1. *CO, 2E,* Page 4-6.

Question #90. Consider a barrier to be like a filter. One or more filters reduces the information flow between the sender and the receiver. NFPA 4.2; 5.2. *CO, 2E,* Pages 23.

Question #91. This is a proactive approach; it involves the activities that help keep fires from occurring. NFPA 5.5. *CO, 2E,* Page 214.

Question #92. Whether volunteer or career, the company officer is responsible for the human resources or staffing along with other resources. NFPA 4.2.1; 4.2.2; 4.2.3. *CO, 2E,* Page 54.

Question #93. This concept is indicated in NFPA 1500 whereas emphasis on accountability and safety during emergency operations has significant emphasis. NFPA 4.7. *CO, 2E,* Page 197.

Question #94. The mission statement declares the vision of the department by setting specific values and setting the focus in which direction the department is moving. A mission statement is like putting up a sign for employees on which direction to go. NFPA 5.2; 4.2. *CO, 2E,* Page 97-98.

Question #95. Fire suppression systems like sprinklers were designed to control the fire growth or even suppress the fire. With this assistance, we have no reported deaths within sprinklered buildings. This system is like giving the fire companies additional personnel for suppression efforts that are housed in each occupancy. NFPA 4.6.2; 4.6.3. *CO, 2E,* Page 253.

Question #96. An organizational chart defines the roles and lines of authority that are important in a paramilitary organizational structure. It is important to know where you fall in the organizational chart as a company officer. NFPA 4.4; 4.6; 5.4; 5.6. *CO, 2E,* Page 47-48.

Question #97. Disciplinary action is another way of improving performance and skills. This lets the subordinate know that their performance is below acceptable standards. NFPA 5.2.2. *CO, 2E,* Page 160.

Question #98. This area is generally left to fire protection engineers and building code officials to assure they meet standards and will protect the structure. NFPA 5.5.1; 4.3.4. *CO, 2E,* Page 223.

Question #99. Goals help you identify where you are going. These take several years often to accomplish. In a sense, this is a strategic road map with the planned stops along the way to reaching the mission. NFPA 4.4. *CO, 2E,* Page 98.

Question #100. Most employees described in the question require some discussion, direction, and supervision. They are beginning to contribute ideas and solutions. NFPA 5.2.1, 4.2.1. *CO, 2E,* Page 128.

Question #101. These are utilized for primary access points barriers to access, utilities, and water supply. Work well for a quick 20-second reference. NFPA 4.6.1. *CO, 2E,* Page 295.

Question #102. Understanding an employee's competency and commitment is important as you mentor and work with these individuals as a supervisor. NFPA 4.2.2. *CO, 2E,* Page 146.

Question #103. Size-up helps start the development of action plans and strategic operations. NFPA 4.6.2. *CO, 2E,* Page 323.

Question #104. Individuals in theory with high-achievement needs tend to work more diligently and want to see it as an achievement. NFPA 5.2.2. *CO, 2E,* Page 125.

Question #105. This is much like a private counseling session to set direction for the subordinate. NFPA 5.2.2. *CO, 2E,* Page 161.

Question #106. The deliberate and apparent process by which one focuses attention on the communications of another is active listening. This is important to the company officer so as to get the full message of the individual presenting the problem and solution. NFPA 4.2.2. *CO, 2E,* Page 24.

Question #107. The company officer gives many oral commands on emergency scenes and they follow customs, rules, and practices of the industry thus making them formal. NFPA 4.2.1. *CO, 2E,* Page 20-23.

Question #108. The command sequence is a three-step process that helps incident commanders manage the incident. NFPA 4.6.2. *CO, 2E,* Page 323-334.

Question #109. This concept of rising heated gases that fill the space and move downward is a natural process called thermal stratification. NFPA 4.7; 4.6.2. *CO, 2E,* Page 257.

Question #110. This area is generally left to fire protection engineers and building code officials to assure they meet standards and will protect the structure. NFPA 5.5.1; 4.3.4. *CO, 2E,* Page 223.

Question #111. . *CO, 2E,* Page 298.

Question #112. This organizational structure follows some of the rules about flat and lean organizations and allows for good communications between the company officer and all of the firefighters on the crew. NFPA 4.2.2. *CO, 2E,* Page 69.

Question #113. Company officers have a status and power. We add this concept to our list of motivators that get employees to do what is asked of them. NFPA 5.2.1. *CO, 2E,* Page 127.

Question #114. Because this type of construction allows long spans without support, it is very popular in large one-story commercial facilities. NFPA 4.6.2. *CO, 2E,* Page 249.

Question #115. This gives the individual a new start and often will allow for changes without embarrassment. NFPA 5.2.2. *CO, 2E,* Page 161.

Question #116. Like playing a scale on a musical instrument where every note is sounded, the scaler principle suggests that every level in the organization is considered in the flow of communications. NFPA 4.4.2. *CO, 2E,* Page 66.

Question #117. The command sequence is a three-step process that helps incident commanders manage the incident. NFPA 4.6.2. *CO, 2E,* Page 323-334.

Question #118. Most organizations have a pyramid structure, with one person in charge, and an increasing number of subordinates at each level as you move downward. The company officer is the first level of supervisor and has a limited number of subordinates for direct supervision, but numbers increase with the size of station and number of incidents. NFPA 4.1.1. *CO, 2E,* Page 62.

Question #119. Each year, over 100 firefighters die in the line of duty. The majority of the fire ground deaths occur when firefighters are advancing hose lines inside the structure. NFPA 4.7.1. *CO, 2E,* Page 197.

Question #120. Understanding fire growth is important as a company officer as you make multiple split-second decisions on tactics and operations based upon the knowledge of fire behavior. NFPA 4.6.2. *CO, 2E,* Page 257.

Question #121. Regardless of whether we are paid or volunteer, we are serving the public and want to be considered and viewed as professionals. NFPA 4.1.1. *CO, 2E,* Page 8.

Question #122. NFPA 4.7. *CO, 2E,* Page 214.

Question #123. The action plan addresses all phases of the incident. NFPA 4.6, 5.6. *CO, 2E,* Page 331.

Question #124. Managing time is an important aspect of a company officer. Time is the one thing that is hard to track for efficiency based upon the tasks performed and the personnel performing them. It is important, however, not to fall into this concept of allowing the prescribed time to dictate the time needed for the event. NFPA5.2.1; 5.2.2; 5.2.3; 4.4.2. *CO, 2E,* Page 105.

Question #125. As a company officer, it is important to understand which style of management practices you must choose. Each situation will require a different style or use of theory. NFPA 5.2.1. *CO, 2E,* Page 81.

Question #126. It is important to know the standard from which the program and requirements are coming. NFPA 1021 1.1. *CO, 2E,* Page 9.

Question #127. Attitude is the core of your performance. Your attitude when positive will reflect a professional and loyal role model who will be able to lead others to meet the department's mission. NFPA 4.1.1. *CO, 2E,* Page 15.

Question #128. The inability to say no or decline the opportunity is the hardest of the time eaters for most. You must remember to tie yourself to goals and focus on priorities. It is important to know when you are overloading your plate. NFPA 5.2.1. *CO, 2E,* Page 106.

Question #129. In the communications process you will be faced with questions for which you may not know the answer. Be up front and honest that you don't know the answer, but work to find the answer. In doing so, you will enhance your knowledge and develop good communications skills. NFPA 4.2.6. *CO, 2E,* Page 26-28.

Question #130. As the company officer gives orders on the emergency scene, the firefighter, who is the receiver of the message, interprets the message and acts from there. NFPA 4.2.2. *CO, 2E,* Page 21.

Question #131. This is everything not covered typically in the capital budget. The operating budget is a general budget that shows specific amounts needed to operate the organization in general form. NFPA 5.4.2; 4.4.3. *CO, 2E,* Page 103.

Question #132. The position of rank affords the right and the power to command. This is the authority you are granted as a company officer. NFPA 4.4.1. *CO, 2E,* Page 99.

Question #133. This is when you accept responsibility for others and their actions as a company officer. NFPA 4.2. *CO, 2E,* Page 62.

Question #134. The incident commander and the company officer are both responsible for these items whereas the initial arriving company officer is the incident commander for at least a few minutes. These priorities are how action plans are formulated. NFPA 5.6.1. *CO, 2E,* Page 320-321.

Question #135. These two pieces are some of the most important parts of what company officers do in a daily function. Human relations are extremely important to the crews and the success of the company officer. NFPA 4.2.2 ; 4.2.3; 5.2.1; 5.2.2. *CO, 2E,* Page 5-7.

Question #136. Classic examples of capital budget items are fire stations, apparatus purchases and any large ticket item. It is important to understand the budgeting process whereas you are managing the money at the first level in protecting the organization's investments. NFPA 4.4.3; 5.4.2. *CO, 2E,* Page 103.

Question #137. All elements of NFPA 1403 and water supply standards must be followed for safety concerns. NFPA 4.7.1. *CO, 2E,* Page 309.

Question #138. This theory is a management theory introduced by Douglas McGregor. This management style in which the manager believes that people dislike work and cannot be trusted was a traditional style of management. NFPA 5.2.1; 5.2.2. *CO, 2E,* Page 81.

Question #139. Reporting to one boss is the concept in this portion. Unity of command is an essential organizational concept. NFPA 4.2.2; 5.2.1. *CO, 2E,* Page 66.

Question #140. Strategy and tactics are set to accomplish the goals for the incident. The priorities are set so as to provide a guide to what areas and tactics need to be taken care of in a sequential order if the situation dictates. NFPA 4.6, 5.6. *CO, 2E,* Page 320.

Question #141. An officer with line authority manages one or more of the functions that are essential for the fire departments mission. When we see an organizational chart, we usually think of the authority one has and in the areas they have it. NFPA 4.4.2; 5.2.1. *CO, 2E,* Page 62.

Question #142. The situation described with an unoccupied structure without power there is little chance of an electrical fire. Arson is motivated by spite, fraud, intimidation, and concealment of a crime etc. NFPA 4.5.1. *CO, 2E,* Page 222.

Question #143. Lightweight construction is engineered to be as strong or stronger than solid components. NFPA 4.6.2. *CO, 2E,* Page 250-251.

Question #144. Life safety education is the first priority in emergency operations. Life safety addresses the safety of occupants and emergency responders. NFPA 4.3.4. *CO, 2E,* Page 222-226.

Question #145. Controlling helps us get to the right place at the right time through the monitoring of efforts of the resources. This is a large part of the company officer's job. NFPA 4.2.1. *CO, 2E,* Page 79.

Question #146. As a leader you will be in one of these roles with different subordinates at the same time. It is important to understand as the individual grows, your role will change. NFPA 5.2.1. *CO, 2E,* Page 151.

Question #147. Rehabilitation is a significant requirement to assure that the personnel working on an emergency scene are physically able to continue working. This is a way to monitor health conditions and potentially recognize signs that could indicate when firefighters are not medically well to continue working. This concept could potentially lower the firefighter death rate in the heart attack area significantly. NFPA 4.7.1. *CO, 2E,* Page 197.

Question #148. Fire investigation is a part of legal proceedings that must follow set steps in order for the materials and information gathered to be used in a legal process. NFPA 4.5.1, 4.5.2, 5.5.2. *CO, 2E,* Page 281.

Question #149. Good communication skills are essential in work and personal life. Formal communications are conducted according to established standards. They tend to follow customs, rules, and practices. NFPA 4.2.2; 5.2.1. *CO, 2E,* Page 20.

Question #150. As company officers, you should note the performance of your personnel during training and actual emergencies. The NFPA standard 1001 allows you to compare their level of knowledge and abilities to a set national job standard. NFPA 4.7.1. *CO, 2E,* Page 308.

Question #151. Total quality management principle is a style often practiced by many company officers as they focus on the organization's continuous improvements and keeping customer satisfaction in mind. NFPA 5.4.5. *CO, 2E,* Page 82.

Question #152. Live fire training provides unique opportunities for developing skills and self-confidence. When acquired structures are used, the building must be carefully inspected and prepared for the training evolution. This standard provides the necessary information to do just that. NFPA 4.7.1. *CO, 2E,* Page 309.

Question #153. Remember that size-up is a good first step in developing an action plan. NFPA 4.6.2, 5.6.1. *CO, 2E,* Page 323-326.

Question #154. Length x Width x Height / 100 = the fire flow for each floor. If the floor is 1/4 involved, then divide that answer by 4 and you would get 2,250 gpm. NFPA 4.6.1. *CO, 2E,* Page 264.

Question #155. Total quality management is based on employee participation and the concept that all must work together to achieve goals. This is true of the fire service, especially on emergency scenes. NFPA 5.2.1. *CO, 2E,* Page 82.

Question #156. Ethics often begin where laws leave off. Ethics have a direct impact on the management of the fire service. These are rules set by the profession. NFPA 4.1.1. *CO, 2E,* Page 89.

Question #157. It is important to understand diversity in the areas we work due to the mixture of people and cultures we can encounter. The answer is based upon understanding the population we serve. NFPA 4.3.1. *CO, 2E,* Page 132.

Question #158. You should understand these systems and how to operate them as a fire officer. NFPA 4.6.1, 5.6.1. *CO, 2E,* Page 253.

Question #159. We utilize these barriers as crutches for not delegating work due to our own insecurity in the delegation process. NFPA 4.2.1; 4.2.2. *CO, 2E,* Page 100.

Question #160. The roles of company officers are many. It is important to know your roles as an officer. An effective company officer fills many roles throughout their careers. Many are done simultaneously. NFPA 4.2. *CO, 2E,* Page 5.

Question #161. Empowerment allows members to have a feeling of ownership in the organization. This is accomplished by getting them involved. NFPA 5.2.1. *CO, 2E,* Page 162.

Question #162. The inability to say no or decline the opportunity is the hardest of the time eaters for most. You must remember to tie yourself to goals and focus on priorities. It is important to know when you are overloading your plate. Procrastination is just the opposite. You do not have a focus on goals or priorities as it relates to work. The continued putting off of this project has you in a position of crunch time for completion. This usually leads to poor work. NFPA 5.2.1. *CO, 2E,* Page 106.

Question #163. This is a mode with which you will never begin. It is a phase that you pass through as you change modes along with strategies and tactics. NFPA 4.6, 5.6. *CO, 2E,* Page 329.

Question #164. The experience of a fire investigator will prove to be beneficial especially with the elements of potential death and potential arson. NFPA 4.5.1; 4.5.2; 5.5.2. *CO, 2E,* Page 277.

Question #165. Our customers are the taxpayers and the employees that work for us. We must address the interior and exterior components of this concept. The three pieces are great guidelines to go by. NFPA 4.3.1; 4.3.3. *CO, 2E,* Page 110.

Question #166. As an organization expands, we assign certain duties to subordinates. It is important that delegation be mutually understood and that authority be provided along with the responsibility. NFPA 4.2.2. *CO, 2E,* Page 67.

Question #167. Principle of scaler organizations is that as you progress with experience and talents, you move upward to higher levels. This is important whereas the company officer is the first level at which individuals accept responsibility for themselves and others. NFPA 4.2. *CO, 2E,* Page 66.

Question #168. As a company officer, it is important to remember to allow in this situation the communications to be a two-way process. NFPA 4.2.2. *CO, 2E,* Page 21.

Question #169. This is the final step in the communications process. NFPA 4.3.3; 4.2.2; 5.2. *CO, 2E,* Page 21.

Question #170. Flashover is a fire phenomena that presents significant risk to firefighters. Fire behavior is a crucial link to how we formulate tactics to accomplish our strategic goals. NFPA 4.7.1, 4.6.2, 5.6.2. *CO, 2E,* Page 258.

Question #171. Stairwells and other penetrations allow for rescue, fire spread, and potential falling hazards for firefighters. These are important to company officers as they work and direct crews in tactical operations. NFPA 4.6.2. *CO, 2E,* Page 258.

Question #172. The company officer's job is as varied as all the activities that the organizations does. However, the company officer is the first-line supervisor and generally the senior representative that the public deals with on a routine basis. NFPA 4.1.1. *CO, 2E,* Page 4.

Question #173. Benchmarks or progression in fire attack is important to the company officer as definitive positions in the incident. NFPA 4.6.3. *CO, 2E,* Page 335.

Question #174. This is the act of guiding the human and physical resources of an organization to attain the organization's objectives. The company officer is the key in this principle whereas they are the ones managing the work force. NFPA 4.2.1; 5.2.1. *CO, 2E,* Page 75.

Question #175. The concept of protecting in place is used in many types of occupancies like high-rise and medical facilities. NFPA 4.6.3, 5.6.1. *CO, 2E,* Page 298.

Question #176. This is the analytical phase of the incident where you gather as much information as possible to help you formulate action plans. NFPA 4.6, 5.6. *CO, 2E,* Page 323-326.

Question #177. Life risk factors are affected by the number of people at risk and their danger and ability to provide for their own safety. There has been no reported fire death in a sprinklered building. NFPA 4.3.1. *CO, 2E,* Page 248.

Question #178. Hygiene factors are easily controlled by company officers and will enhance the employee's work performance. NFPA 4.7; 5.7 ; 5.2.1. *CO, 2E,* Page 125.

Question #179. Supervisors should be sure that all employees are fairly treated and represented in all department activities. NFPA 5.2.1; 4.7.1; 5.2.2. *CO, 2E,* Page 131.

Question #180. The number of people you can effectively supervise varies, of course, based on many factors, but generally for our purposes the number is between four and seven. NFPA 5.2.1; 5.6.1. *CO, 2E,* Page 67.

Question #181. In writing, it is important to remember who is going to read what you wrote. This way, you will write to have effective communications with no barriers. NFPA 4.3.2. *CO, 2E,* Page 28.

Question #182. Dynamic and effective leaders make their style fit the situation. There is importance in utilizing the different levels based on the individuals and the tasks. NFPA 4.2.3; 5.2.2. *CO, 2E,* Page 128.

Question #183. The third step in the management process, commanding, involves using the talents of others, giving them directions, and setting them to work. NFPA 5.6.1. *CO, 2E,* Page 78.

Question #184. The role of the company officer is to manage personnel at the crew level. Performance and safety are key items for which the officer is responsible. NFPA 4.2.1. *CO, 2E,* Page 4.

Question #185. Leading others is the company officer's principle job. The capabilities, efficiency, and morals of the company are direct reflections of the company officer's leadership abilities. NFPA 5.2; 4.2. *CO, 2E,* Page 5.

Question #186. Locations where there are unusual hazards or where an incident would overload the department's resources are examples of target hazards. Nursing facilities and retirement centers are high life hazard complexes. NFPA 4.6.1. *CO, 2E,* Page 293.

Question #187. Safety comes from the elimination of unsafe acts or equipment. It is important as a company officer to recognize the issues and the requirements and standards that support those issues. NFPA 4.7.1. *CO, 2E,* Page 182-183.

Question #188. It is important to understand the causes of injuries and death of firefighters to take a proactive approach to reduce these numbers. NFPA 4.7.1; 4.7.2. *CO, 2E,* Page 179.

Question #189. Physical fitness lowers injury and death rates by conditioning firefighters for the job. It is important as an officer to know what standards are in place for safety. NFPA 4.7.1; 4.7.2. *CO, 2E,* Page 183.

Question #190. Planning covers everything from the next hour to the next decade. NFPA 4.2.3. *CO, 2E,* Page 77.

Question #191. Knowing the leading causes of fire as being careless smoking and the evidence presented, you would tend to lean toward that answer based upon fire behavior and the time of day coupled with the ashtray finding and the smoking resident. However, the accidental causes of fires are lead by heating equipment. The question brings forth temperature, time of year, and potentials. NFPA 4.5.1. *CO, 2E,* Page 221.

Question #192. Action plans for fire attack are based upon fire behavior and the building construction to determine the appropriate tactics. NFPA 4.6.1. *CO, 2E,* Page 249.

Question #193. NFPA 1500 provides policies for managing events to include accountability. NFPA 4.7.1; 4.6.3; 5.6.1. *CO, 2E,* Page 197.

Question #194. The first step in the management process is planning. Planning can be for short-, long-, or mid-term time frames. NFPA 4.2.2. *CO, 2E,* Page 77.

Question #195. It is important to understand the requirements to which you are certifying as a company officer. NFPA 4.1. *CO, 2E,* Page 9-11.

Question #196. A goal is a target or object by which achievement can be measured. As company officers, we set departmental, company, and individual goals for ourselves and others daily. NFPA 4.2.3. *CO, 2E,* Page 98.

Question #197. The priorities for any incident are as follows in order: life safety, incident stabilization, and property conservation. NFPA 4.6.2. *CO, 2E,* Page 320.

Question #198. NFPA 1500 states every department should have an individual assigned these duties. NFPA 4.7.1; 5.7.1. *CO, 2E,* Page 184.

Question #199. All high-rise buildings are made of fire-resistive construction. NFPA 4.6.2. *CO, 2E,* Page 248.

Question #200. The first step in the management process is planning and the effectiveness of a company officer is the ability to make and keep plans moving forward. NFPA 5.4.5. *CO, 2E,* Page 77.

Glossary

access factors an assessment of the department's access to and into a building

accidental refers to those fires that are the result of unplanned or unintentional events

accountability being responsible for one's personal activities; in the organizational context, accountability includes being responsible for the actions of one's subordinates

action plan an organized course of action that addresses all phases of incident control within a specified period

active listening the deliberate and apparent process by which one focuses attention on the communications of another

administrative law a body of law created by administrative agencies in the form of rules, regulations, orders, and court decisions

administrative process a body of law that creates public regulatory agencies and defines their powers and duties

administrative search warrant a written order issued by a court specifying the place to be searched and the reason for the search

agency shop an arrangement whereby employees are not required to join a union, but are required to pay a service charge for representation

arbitration the process by which contract disputes are resolved with a decision by a third party

area of refuge portion of a structure that is relatively safe from fire and the products of combustion

arson a legal term denoting deliberate and unlawful burning of property

attack role a situation in which the first-arriving officer elects to take immediate action and to pass command on to another officer

authority the right and power to command

automatic fire protection sprinkler self-operating thermosensitive device that releases a spray of water over a designed area to control or extinguish a fire

backdraft explosion a type of explosion caused by a sudden influx of air into a mixture of burning gases that have been heated to the ignition temperature of at least one

barrier an obstacle; in communications, a barrier prevents the message from being understood by the receiver

benchmarks significant points in the emergency event usually marking the accomplishment of one of the three incident priorities: life safety, incident stabilization, or property conservation

British thermal unit the amount of heat required to raise the temperature of one pound of water one degree Fahrenheit

budget a financial plan for an individual or organization

building code law or regulation that establishes minimum requirements for the design and construction of buildings

capital budget a financial plan to purchase high-dollar items that have a life expectancy of more than 1 year

certificates a document serving as evidence of the comple tion of an educational or training program; term also describes a document issued to an individual or company as a fire prevention tool

certification a document that attests that a person has demon strated the knowledge and skills necessary to function in a particular craft or trade

closed shop a term in labor relations denoting that union membership is a condition of employment

coach a person who helps another develop a skill

codes a systematic arrangement of a body of rules

command role a situation in which the first-arriving officer takes command until relieved by a senior officer

commanding the third step in the management process, commanding, involves using the talents of others, giving them directions and setting them to work

community consequences an assessment of the consequences on the community, which includes the people, their property, and the environment

complaint an expression of discontent

confinement an activity required to prevent fire from extending to an uninvolved area or another structure

conflict a disagreement, quarrel, or struggle between two individuals or groups

consensus document the result of general agreement among members who contribute to the document

consulting seeking advice or getting information from another; as a leadership style it implies that the leader seeks ideas and allows contributions to the decision-making process

controlling the fifth step in the management process, controlling, involves monitoring the process to ensure that the work is accomplishing the intended goals and objectives, and taking corrective action when it is not

coordinating the fourth step in the management process, coordinating, involves the manager's controlling the efforts of others

counseling one of several leadership tools that focuses on improving member performance

customer service service to people in the community

defensive mode actions intended to control a fire by limiting its spread to a defined area

delegate to grant to another a part of one's authority or power

delegating sharing work, authority, and responsibility with another; as a leadership style, delegating implies the most generous sharing of the officer's leadership role

delegation the act of assigning duties to subordinates

demotion reduction of a member to a lower grade

Dillon's Rule a legal ruling from Chief Justice Dillon of the Iowa Supreme Court whereby local governments have only those powers expressly granted by charter or statute

directing controlling a course of action; as a leadership style, it is characterized by an authoritarian approach

disciplinary action an administrative process whereby a member is punished for not conforming to the organizational rules or regulations

discipline a system of rules and regulations

diversity a quality of being diverse, different, or not all alike

emergency incident any situation to which a fire depart

ment or other emergency response organization responds to deliver emergency services

empowerment to give authority or power to another

energy the capacity to do work; in the fire triangle, heat represents energy

environmental factors factors like weather that impact on firefighting operations

Equal Employment Opportunity Commission (EEOC) federal government agency charged with administering laws related to nondiscrimination on the basis of race, color, religion, sex, age, or national origin

ethics a system of values; a standard of conduct

expert power a recognition of authority by virtue of an individual's skill or knowledge

fact finding a collective bargaining process; the fact finder gathers information and makes recommendations

Fayol's bridge organizational principle that recognizes the practical necessity for horizontal as well as vertical communications within an organization

feedback reaction to a process that may alter or reinforce that process

finance section that part of the incident management system that is responsible for facilitating the procurement of resources and for tracking the costs associated with such procurement

fire behavior the science of the phenomena and consequences of fire

fire control activities associated with confining and extinguishing a fire

fire extension the movement of fire from one area to another

fire extinguishment activities associated with putting out all visible fires

fire load stuff that will burn

fire prevention code legal document that sets forth the requirements for life safety and property protection in the event of fire, explosion, or similar emergency

fire prevention action taken to prevent a fire from occur ring, or if one does occur, to minimize the loss

fire suppression action taken to control and extinguish a fire

fire-resistive construction a type of building construction in which the structural components are noncombustible and protected from fire

first-level supervisors the first supervisory rank in an organization

flashover a dramatic event in a room fire that rapidly leads to full involvement of all combustible materials present

floor plan a bird's-eye view of the structure with the roof removed showing walls, doors, stairs, and so on

free-burning phase second phase of fire growth, has sufficient fuel and oxygen to allow for continued fire growth

fuel load the expected maximum amount of combustible material in a given fire area

goal a target or other object by which achievement can be measured; in the context of management, a goal helps define purpose and mission

grievance procedure formal process for handling disputed issues between member and employer; where a union contract is in place, the grievance procedure is part of the contract

grievance a formal dispute between member and employer over some condition of work

gripe the least severe form of discontent

harassment to disturb, torment, or pester

health and safety officer a person assigned as the manager of the department's health and safety program

heat of combustion the amount of heat given off by a particular substance during the combustion process

heavy timber construction a type of building construction in which the exterior walls are usually made of masonry, and therefore, noncombustible

high-achievement needs according to the needs theory of motivation, individuals with high-achievement needs accept challenges and work diligently

high-affiliation needs according to the needs theory of motivation, individuals with high-affiliation needs desire to be accepted by others

hygiene factors as used by Frederick Herzberg, hygiene factors keep people satisfied with their work environment

identification power a recognition of authority by virtue of the other individual's character or trust

ignition temperature the minimum temperature to which a substance must be heated to start combustion after an ignition source is introduced

incident command system a tool for dealing with emergencies of all kinds

incident commander the person in overall command of an incident

incident management system an organized system of roles, responsibilities, and standard operating procedures used to manage an emergency operation

incident safety officer as part of the IMS, that person responsible for monitoring and assessing safety hazards and ensuring personnel safety

incipient phase first stage of fire growth, limited to the material originally ignited

information officer as part of the IMS, that person who acts as the contact between the IC and the news media, responsible for gathering and releasing incident information

initial report a vivid but brief description of the on-scene conditions relevant to the emergency

in-service company inspections using suppression companies to conduct fire safety inspections in selected occupancies, usually within the company's first-due assignment area

intentionally set The term intentionally set is now used by NFPA and USFA to describe fires that were deliberately set.

internal customers the members of the organization

latent heat of vaporization the amount of heat required to convert a substance from a liquid to a vapor

leadership the personal actions of managers and supervisors to get team members to carry out certain actions

legitimate power a recognition of authority derived from the government or other appointing agency

liaison officer as part of the incident management system, the contact between the incident commander and agencies not represented in the incident command structure

licenses formal permission from an authority to participate in an activity

life risk factors the number of people in danger, the immediacy of their danger, and their ability to provide for their own safety
life safety education the first priority in emergency operations, life safety addresses the safety of occupants and emergency responders
life safety the first priority during all emergency operations that addresses the safety of occupants and emergency responders
line authority a characteristic of organizational structures denoting the relationship between supervisors and subordinates
line functions refer to those activities that provide emergency services
line-item budgeting collecting similar items into a single account and presenting them on one line in a budget document
lock box a locked container on the premises that can be opened by the fire department containing preplanning information, material safety data sheets, names and phone numbers of key personnel, and keys that allow emergency responders access to the property
logistics section that part of the incident management system that provides equipment, services, material, and other resources to support the response to an incident

management by exception management approach whereby attention is focused only on the exceptional situations when performance expectations are not being met
management the accomplishment of the organization's goals by utilizing the resources available
Maslow's Hierarchy of Needs a five-tiered representation of human needs developed by Abraham Maslow
mediation the process by which contract disputes are resolved with a facilitator
message in communications, the message is the information being sent to another
minimum standards as used in codes and standards, it indicates the least or lowest accepted level of attainment that is acceptable
mission statement a formal document indicating the focus and values for an organization
model representation or example of something
motivators factors that are regarded as work incentives such as recognition and the opportunity to achieve personal goals
motive the goal or object of one's actions; the reason one sets a fire
mutual aid assistance provided by another fire department or agency

noncombustible construction a type of building construction in which the structural elements are noncombustible or limited combustible

objective something that one's efforts are intended to accomplish
occupancy factors an assessment of the risks associated with a particular structure based on the contents and activities therein
occupant services sector that part of the incident management system that focuses on the needs of the citizens who are directly or indirectly affected by an incident
offensive mode firefighting operations that make a direct attack on a fire for purposes of control and extinguishment
open shop a labor arrangement in which there is no requirement for the employee to join a union
operating budget a financial plan to acquire the goods and services needed to run an organization for a specific period of time, usually 1 year
operations section division in the incident management system that oversees the functions directly involved in rescue, fire suppression, or other activities within the mission of an organization
oral reprimand the first step in a formal disciplinary process
ordinary construction a type of building construction in which the exterior walls are usually made of masonry, and therefore, noncombustible: the interior structural members may be either combustible or noncombustible
organization a group of people working together to accomplish a task
organizing the second step in the management process, organizing, involves bringing together and arranging the essential resources to get a job done
overhaul searching the fire scene for possible hidden fires or sparks that may rekindle
oxidation a chemical reaction in which oxygen combines with other substances causing fire, explosions, and rust

permits fire prevention tool required where there is potential for life loss, or where there are hazardous materials or hazardous processes
personnel accountability the tracking of personnel as to location and activity during an emergency event
physical factors an assessment of the conditions relevant to population, area, topography, and valuation of a given area
planning section that part of the incident management system that focuses on the collection, evaluation, dissemination, and use of information (information management) to support the incident command structure
planning the first step in the management process, planning, involves looking into the future and determining objectives
plot plan a bird's-eye view of a property showing existing structures for the purpose of pre-incident planning, such as primary access points, barriers to access, utilities, water supply, and so on
policy formal statement that defines a course or method of action
power the command or control over others, status
pre-incident planning preparing for operations as the scene of a given hazard or occupance
pre-incident survey the fact-finding part of the pre-incident planning process in which the facility is visited to gather information regarding the building and its contents
probable cause a reasonable cause for belief in the existence of facts
procedure a defined course of action
program budget the expenses and possible income related to the delivery of a specific program within an organization
property conservation the efforts to reduce primary and secondary (as a result of firefighting operations) damage
property risk factors an assessment of the value and hazards associated with property that is at risk
public fire department a part of local government

punishment power a recognition of authority by virtue of the supervisor's ability to administer punishment

Quick Access Prefire Plan a document that provides emergency responders with vital information pertaining to a particular occupancy

rapid intervention team (RIT) a team or company of emergency personnel kept immediately available for the potential rescue of other emergency responders
receiver in communications, the receiver is the intended recipient of the message
rehabilitation as applies to firefighting personnel, an opportunity to take a short break from firefighting duties to rest, cool off, and replenish liquids
resource factors an assessment of the resources available to mitigate a given situation
responsibility being accountable for actions and activities; having a moral and perhaps legal obligation to carry out certain activities
reward power a recognition of authority by virtue of the supervisor's ability to give recognition
rollover reignition of gases that have risen and encountered fresh air, and thus a new supply of oxygen

scaler principle organizational concept that refers to the interrupted series of steps or layers within an organization
search warrant legal writ issued by a judge, magistrate, or other legal officer that directs certain law enforcement officers to conduct a search of certain property for certain things or persons, and if found, to bring them to court
sector, section, division, group these terms are tactical-level management groups that command companies
sender a part of the communications process, the sender transmits a thought or message to the receiver
size-up mental assessment of the situation; gathering and a nalyzing information that is critical to the outcome of an event
smoldering phase third stage of fire growth; once the oxygen has been reduced, visible fire diminishes
span of control organizational principle that addresses the number of personnel a supervisor can effectively manage
specific heat the heat-absorbing capacity of a substance
specifications spell out in detail the type of construction or the materials to be used
staff assistants those designated to aid the IC in fulfilling the IC's responsibilities, typically, an incident safety officer, an information officer, and a liaison officer.
staff functions refer to those activities that support those providing emergency services
standard of performance a defined level of accomplishment or achievement
standard operating procedure (SOP) an organized directive that establishes a standard course of action
standard a rule for measuring or a model to be followed
standpipe systems plumbing installed in a building or other structure to facilitate firefighting operations
strategy sets broad goals and outlines the overall plan to control the incident
structural factors an assessment of the age, condition, and structure type of a building, and the proximity of exposures
supporting as a leadership style, the supporting process involves open and continuous communications and a sharing in the decision-making process
survival factors an assessment of the safety hazards for both civilians and firefighters in a particular occupancy
suspension disciplinary action in which the member is relieved from duties, possibly with partial or complete loss of pay.

tactics various maneuvers that can be used to achieve a strategy while fighting a fire or dealing with a similar emergency
target hazards locations where there are unusual hazards, or where an incident would likely overload the department's resources, or where there is a need for interagency cooperation to mitigate the hazard
tasks the duties and activities performed by individuals, companies, or teams that lead to successful accomplishments of assigned tactics
termination the final step in the disciplinary process or the incident command process
theoretical fire flow the water flow requirements expressed in gallons per minute needed to control a fire in a given area
Theory X management style in which the manager believes that people dislike work and cannot be trusted
Theory Y management style in which the manager believes that people like work and can be trusted
Theory Z management style in which the manager believes that people not only like to work and can be trusted, but that they want to be collectively involved in the management process and recognized when successful
thermal stratification rising of hotter gases in an enclosed space
total quality management focus of the organization on continuous improvement geared to customer satisfaction
transfer a step in the disciplinary process that provides the member a fresh start in another venue
transitional mode the critical process of shifting from the offensive mode or from the defensive to the offensive

union shop a term used to describe the situation in which an employee must agree to join the union after a specified period of time, usually 30 days after employment
unit of command the organizational principle whereby there is only one boss

vented opened to the atmosphere by the fire burning through windows or walls through which heat and fire by-products are released, and through which fresh air may enter
ventilation a systematic process to enhance the removal of smoke and fire by-products and the entry of cooler air to facilitate rescue and fire-fighting operations
vision an imaginary concept, usually favorable, of the result of an effort

woodframe construction a type of building construction in which the entire structure is made of wood or other combustible material
written reprimand documents unsatisfactory performance and specifies the corrective action expected; usually follows an oral reprimand